EMBLEMS OF
EXPLORATION

Logos of the NACA and NASA

Monographs in Aerospace History, no. 56

Joseph R. Chambers
Mark A. Chambers

National Aeronautics and Space Administration

Office of Communications
NASA History Program Office
Washington, DC 20546

NASA SP-2015-4556

Library of Congress Cataloging-in-Publication Data

Chambers, Joseph R.
    Emblems of exploration : logos of the NACA and NASA / by Joseph R. Chambers and Mark A. Chambers, National Aeronautics and Space Administration, Office of Communications, Public Outreach Division, History Program Office.
        pages cm. — (Monographs in aerospace history ; #56)
    Includes bibliographical references.
1. United States. National Aeronautics and Space Administration—Insignia. 2. United States. National Aeronautics and Space Administration—Medals. I. Chambers, Mark A. II. Title.
    TL521.312.C438 2015
    629.402'7—dc23
                                                    2015019236

This publication is available as a free download at *http://www.nasa.gov/ebooks*.

ISBN 978-1-62683-028-8

Contents

Preface

In 1992, Jerry C. South, Jr., a NASA researcher at the NASA Langley Research Center in Hampton, Virginia, posed a challenge to A. Gary Price, Head of the Office of External Affairs at Langley. The new administrator of NASA, Daniel S. Goldin, had recently made a decision to reinstitute the well-known "meatball" insignia as part of his effort to revitalize the Agency. South had knowledge of some aspects of the origin of the emblem that had been designed 34 years earlier in 1958, and he pointed out that the details of the conception and history of the logo had never been documented. Since Price's organization was the designated Langley focal point for such historical information at the time, South suggested that the topic would be an excellent project for External Affairs.[1] Mark Chambers, the coauthor of this publication, worked for Price and was assigned the task of researching the genesis of the emblem, including interviewing key individuals known to be essential to the NASA meatball story.

In the summer of 1992, Chambers interviewed several key participants involved in the design process, reviewed correspondence, studied documents and technical reports, and obtained relevant material from the NASA photographic archives at Langley. The story that emerged from his efforts was reported to Price and his organization, which in turn sent samples of some of the documented results to NASA Headquarters for official review.[2] Unfortunately, the results of the study were never published.

In 2012, Chambers' father, Joseph R. Chambers, discovered the 20-year-old notes that had been accumulated by Mark and decided to pursue the story with the intention of formally documenting the details to the fullest extent possible. By that time, many of the key participants of the events that had occurred almost 55 years earlier had passed away, and much more research was required to fill in the blanks.

After the completion of additional research and new interviews, a lecture on the evolution of the NASA seal and insignia was presented to Langley staff and

---

1   South, J. C., Jr., *Meatball Logo Based on Wind Tunnel Model, The Researcher News*, Langley's in-house newsletter, NASA Langley Research Center, 14 August 1992, p. 2.

2   Chambers, Mark A., "History of the Red 'V' in the NASA Meatball," NASA History Office File 4542, 31 July 1992.

the public in 2013, to good reception.[3] The presentation was then followed by overview articles in *News & Notes,* the NASA History Program Office's quarterly newsletter.[4] As interest in the topic continued to grow, and with the encouragement of the NASA Headquarters History Program Office, the authors expanded the story of the NASA emblems to include additional research and discussion of the logos used by NASA's predecessor, the National Advisory Committee for Aeronautics (NACA).

The expansion of the effort to document the complete account of the design, evolution, and applications of the NACA and NASA emblems was deemed particularly appropriate in recognition of the 100th anniversary of the NACA/NASA in 2015. The centennial celebration includes numerous historical reviews, symposia, and media coverage that will benefit from the material herein.

Research on the NACA emblems was especially challenging. The history and evolution of the symbols had not been documented prior to the current research. Tracking down official documents and photographs for the period 1915–1958 required extensive investigation of current archives at NASA Centers and trips to the National Archives and Records Administration (NARA) at College Park, Maryland, where many of the documents and photographs of the NACA laboratories are stored. Unfortunately, some of

the photographs showing early applications of NACA logos have deteriorated beyond repair with age or have disappeared from archival collections. On the other hand, the majority of textual records at NARA have been preserved in excellent physical condition and thus provided invaluable information for this effort.

As might be expected, surprises occurred in the review of the NACA data. For example, it was found that the NACA operated for its first 25 years without an official insignia and for almost 40 years without a seal during an age noted for heraldry and organizational emblems. The historical data gathered to date do not permit identification of the designer of the original NACA logo. Fortunately, archived information did permit identification of precise dates and descriptions of changes that occurred during the evolution of the NACA insignia and seal.

Documentation of the design, evolution, and applications of NASA emblems proved to be a less difficult task. Key individuals and agencies involved in the design process were identified, the symbolism of certain elements was determined, changes that occurred over the years were documented, and examples of applications were selected. Research on the NASA logos also produced unexpected results and findings. For example, the background behind the design of the red "slash" exhibited in the NASA insignia revealed an undocumented association with previously classified research activities. Also surprising was the unwillingness of organizations—especially flight research operations—to follow directives regarding the application and modification of logos.

The material presented herein is organized chronologically and covers the subject from the first days of the NACA in 1915 to the current-day situation of NASA. Inputs from historical archivists at the current NASA Centers have contributed a significant amount of material on the logos of the NACA and NASA at their

---

3   Chambers, Joseph R., "Wings, Meatballs, Worms and Swooshes," Lecture presented at NASA Langley Research Center Colloquium Series, 9 July 2013, *http://www. youtube.com/watch?v=uLRMNNQiE0Q&feature=youtu.be* (accessed on 11 August 2014).

4   Joseph Chambers, "The NASA Seal and Insignia, Part 1," *News & Notes* of the NASA History Program Office, 30, no. 2 (2013), pp. 11–15; and "The NASA Seal and Insignia, Part 2," *News & Notes*, 30, no. 3 (2013), pp. 1–4. Back issues of *News & Notes* are available at *http://history.nasa. gov/histnews.htm.*

specific Centers. Particularly valuable information on the chronology of the NACA and NASA logos and the participation of the U.S. Army is also discussed.

This publication has two primary objectives regarding the NACA and NASA logos. The first objective is to provide, to the extent possible, detailed information about the designers and design processes for the emblems. The second objective of the discussions is to briefly illustrate the applications of these respected and admired insignias and seals within the cultures of each agency. For this task, photographs and descriptions are used to exemplify applications to buildings; equipment; aircraft and spacecraft; correspondence and documents; and personal memorabilia such as pins, awards, and retirement plaques.

In view of the hundreds of thousands of photographs that might have been used to illustrate applications of the NACA and NASA logos, the authors took on the imposing task of reducing the candidates to an acceptable number of examples for this publication. Within this necessary constraint, we chose photographs that illustrate especially important and interesting activities. We humbly apologize to all for omission of other favorite projects and facilities.

Over the course of their histories, the NACA and NASA have developed a wide variety of emblems representing each agency's illustrious exploration of aerospace missions. Hundreds of individual mission emblems and patches came about with the beginning of human spaceflight activities that will not be covered herein. This publication concentrates on the rich and interesting history of the conception and implementation of the world-famous NACA and NASA seals and insignias that have been displayed for decades on aeronautics and space research vehicles and facilities, as well as those proudly worn by flight research pilots, astronauts, and the dedicated employees of these two world-class organizations.

Discussions in each chapter refer to certain NACA laboratories and NASA Centers as they existed in specific time periods before and after the establishment of NASA in 1958. For the reader's information, the locations and name changes of the sites through the years are as follows:

Hampton, Virginia: Langley Memorial Aeronautical Laboratory (LMAL 1920–1948); Langley Aeronautical Laboratory (LAL 1948–1958); Langley Research Center (LaRC 1958–present)

Moffett Field, California: Ames Aeronautical Laboratory (AAL 1939–1958); Ames Research Center (ARC 1958–present)

Cleveland, Ohio: Aircraft Engine Research Laboratory (AERL 1942–1947); Flight Propulsion Research Laboratory (FPRL 1947–1948); Lewis Flight Propulsion Laboratory (LFPL 1948–1958); Lewis Research Center (LRC 1958–1999); John H. Glenn Research Center (GRC 1999–present)

Edwards, California: Muroc Flight Test Unit (1947–1949); High-Speed Flight Research Station (HSFRS 1949–1954); High-Speed Flight Station (1954–1959); Flight Research Center (FRC 1959–1976); Dryden Flight Research Center (DFRC 1976–1981); Ames-Dryden Flight Research Facility (1981–1994); Dryden Flight Research Center (DFRC 1994–2014); Armstrong Flight Research Center (AFRC 2014–present)

Wallops Island, Virginia: Pilotless Aircraft Research Station (PARS 1945–1958); Wallops Station (WS 1958–1974); Wallops Flight Center

(WFC 1974–1981); Wallops Flight Facility (WFF 1981–present)

Houston, Texas: Manned Spacecraft Center (MSC 1963–1973); Johnson Space Center (JSC 1973–present)

Merritt Island, Florida: Launch Operations Center (LOC 1962–1963); Kennedy Space Center (KSC 1963–present)

Acknowledgments

Special thanks go to William P. Barry, NASA Chief Historian, for providing the encouragement and mechanism for this undertaking. Within the NASA Headquarters History Program Office, Stephen J. Garber served as contract manager for NASA, and Jane H. Odom and Colin A. Fries provided valuable documents from the archival collection at Headquarters. At the Langley Research Center, Mary E. Gainer provided photographs, documents, and support within her responsibilities for preserving Langley's artifacts and documenting its history via the site *http://crgis.ndc.nasa.gov/historic/Larc*, which is available for public viewing. Her intern, James R. Baldwin, provided photographs of current Langley buildings. H. Garland Gouger, Susan K. Miller, and the staff of the Langley Technical Library provided invaluable documents and information, especially for the early NACA days. Langley's photo archivist Teresa L. Hornbuckle contributed personal research efforts and access to previously unpublished photographs in the Langley collection, and Gail S. Langevin provided access to the Langley Historical Archive Collection. Katrina L. Young of Langley's Awards Office provided extensive documentation on NASA medals and other awards. Langley employees contributing artifacts included Tracey E. Redman, Benjamin A. Goddin, and James T. Berry. Graphics specialists Stanley H. Husch, William B. Kluge, and James H. Cato provided valuable information regarding current NASA graphics standards, aircraft insignia, and additional graphics support.

We are deeply appreciative of the support contributed by individuals at other NASA Centers. Thanks are extended to historian Robert S. Arrighi of the NASA Glenn Research Center, who provided unique photographs and documents on the career and personal life of James J. Modarelli as well as examples of applications of the logos at the Center. At the NASA Ames Research Center, historian Glenn E. Bugos, photo archivist Lana L. Albaugh, and archivist April D. Gage contributed photographs and information on design efforts for the NACA and NASA emblems and examples of applications of logos at Ames. Glenn also provided documents obtained during his visits to the National Archives and Records Administration (NARA) facility at San Bruno, California. Historians Christian Gelzer and Peter W. Merlin of the NASA Armstrong Flight Research Center provided valuable comments and photographs that added great value to the effort. Special thanks to Peter for his substantial contribution of photographs and comments.

Many active and retired NASA personnel provided the foundation for the technical discussion and personal anecdotes presented in this work. Sincere thanks to Langley retirees Jerry South, Jr.; A. Gary Price; Roy V. Harris, Jr., P. Kenneth Pierpont; Jack W. Crenshaw; Harry W. Carlson; A. Warner Robins; and Owen G. Morris; Ames retirees Victor Peterson, Leroy L. Presley, Vernon J. Rossow, and William C. Pitts; and Johnson Space Center retirees Roger Zwieg and Robert M. Payne. Helene Katzen's permission to use a photograph of her husband Elliott Katzen is also appreciated.

The staff of the NARA facility at College Park, Maryland, was of tremendous assistance, providing access and guidance for finding original documents and photographs from its NACA collection. Particular thanks go to NARA archivists David A. Pfeiffer of the Textual Reference Branch and Holly Reed of the Still Pictures Branch. Brian Nicklas of the Smithsonian National Air and Space Museum also provided insightful comments and photographs. Aviation enthusiast Jennings Heilig contributed graphic support. Special thanks to Keith Yoreg and his mother Janette Davis for their gracious contribution of the photograph of Orville Wright's badge and to James J. Modarelli, Jr., for providing materials and anecdotes on his father's legendary career. Scott Sacknoff, publisher of *Quest: The History of Spaceflight Quarterly*, also provided graphics and valuable comments.

Charles V. Mugno, Director of the Army Institute of Heraldry, and Paul J. Tuohig, senior research analyst, shared detailed records of the interactions of their agency with the NACA and NASA during the design evolution of the seals and insignia. Their inputs were a key contribution to the success of this effort.

Thanks to the anonymous reviewers who examined the draft for accuracy, content, and readability. Their comments and suggested changes resulted in a greatly enhanced publication. Finally, the authors wish to thank Yvette Smith of the NASA History Program Office as well as Chinenye Okparanta, Jennifer Way, and Michele Ostovar of NASA's Communications Support Services Center for their outstanding work in editing and layout.

The authors were honored to have interviewed principal participants Robert T. Jones and James J. Modarelli before their passing. Although the details of memories of events that had occurred over 34 years earlier at the time of their interviews had begun to fade, the discussions and information they provided were priceless.

CHAPTER 1

Wings of Eagles, 1915–1940

Beginnings

Following the first controlled heavier-than-air flight by the Wright brothers in 1903, a relative lack of interest in further development and applications of this radical new capability existed within the United States. Meanwhile, visionary individuals in Europe took the lead in furthering technologies and accelerating experiences with applications of new flying machines. When World War I began in 1914, France had 1,400 airplanes; Germany had 1,000; Russia had 800; Great Britain had 400; and the United States had only 23.[1] The loss of U.S. momentum and leadership in human flight did not, however, go unnoticed by a small group of American scientists, politicians, and public figures. In 1913, Charles D. Walcott, the new Secretary of the Smithsonian Institution, attempted to stimulate Congress into creating a new agency for aeronautical research and oversight.

Despite the fact that President Woodrow Wilson did not favor Walcott's proposal, the untiring efforts of Walcott and a small group of scientists and military officers finally moved Congress to take action. Congress created a rider to a 1915 Naval Appropriations Act to create a new aeronautical advisory committee to organize and direct aeronautical research and development for the nation. The Advisory Committee, which was created on 3 March 1915, reported directly to the President, who personally appointed its members with no salaries. The legislation gave birth to a new agency initially named the Advisory Committee for Aeronautics, similar to an established advisory group in England. The Committee originally consisted of 12 members, with 2 members from the Army, 2 from the Navy, and 1 each from the Smithsonian Institution, the National Bureau of Standards, and the Weather Bureau. An additional 5 members were selected from the engineering and scientific communities. At its first meeting, the group proclaimed itself the National Advisory Committee for Aeronautics (NACA).

---

1 J. C. Hunsaker, "Forty Years of Aeronautical Research," *Smithsonian Report for 1955*, p. 243.

The Langley Memorial Aeronautical Laboratory

The rider that established the NACA also included the potential for future establishment of an aeronautical research laboratory. In 1916, Walcott and Army and Navy leaders discussed a proposal to establish a joint Army-Navy-NACA experimental field and proving ground. After extensive debates over potential sites for the research laboratory, the War Department procured 1,650 acres of land in 1917 near Hampton, Virginia, for the combined use of the Army, Navy, and the NACA in aeronautical operations and research.[2] The site was named Langley Field in honor of early flight pioneer Samuel Pierpont Langley. The NACA began construction of the first civilian research laboratory there in 1917. Originally known as the Field Station of the Committee, it was renamed on 11 June 1920 the Langley Memorial Aeronautical Laboratory (LMAL).[3]

World-famous architect Albert Kahn of Detroit, Michigan, was chosen by the Army to design the general layout of the property, the flying field, and the Army Air Service buildings at Langley Field in 1917. The construction process proved to be a tremendous challenge because of the isolated location—complete with swarms of mosquitoes, outbreaks of influenza, inadequate housing—and poor relations between the Army and the NACA. Delays in construction of facilities and bickering between the intended organizations led to the Army transferring its research organization to McCook Field in Ohio, while retaining an aviation training mission at Langley Field. The Navy also abandoned its role at the new site and moved its testing of seaplanes to Norfolk, Virginia.

The First Wings

Although Kahn had overall leadership for the detailed design of the Army buildings at Langley, the NACA contracted with the architectural firm of Donn & Deming of Washington, DC, for the design of its laboratory building on 13 June 1917.[4] Construction efforts by the J. G. White Engineering Corporation for the NACA laboratory building began on 17 July 1917 with excavation. The building was completed in 1918 and construction of the first NACA wind tunnel began.[5] The LMAL officially dedicated its laboratory in conjunction with the completion of its atmospheric wind tunnel on 11 June 1920. By 1924, the NACA buildings included the original laboratory building (by then an administration building), an atmospheric wind tunnel with a 5-foot test section, the world's first variable-density wind tunnel, two temporary buildings serving as dynamometer labs, and a small airplane hangar at the Langley flight line.

William I. Deming and his team designed the LMAL building with Doric columns, a projecting cornice, and limestone door surrounds. An engraving of a circular symbol containing stylized versions of the

---

2   The NACA's initial share of the property was only 5.8 acres but as the research facilities multiplied in ensuing years, the War Department released more land and, by 1939, the NACA occupied 28 acres.

3   Letter from John F. Victory, Assistant Secretary of the NACA to Langley Memorial Aeronautical Laboratory, Subject: Change of Laboratory Name. 16 June 1920. (Record Group 255, Textual Reference Branch, NARA College Park, MD)

4   Letter from John F. Victory, NACA Special Disbursing Agent to Donn & Deming, Architects of Washington, DC, dated 13 June 1917. (Record Group 255, NARA Textual Reference Branch, College Park, MD)

5   Donn & Deming was also responsible for the design of the building housing the wind tunnel.

FIGURE 1-1.

The Langley Memorial Aeronautical Laboratory in 1924. The building in the foreground housed administration and services. The building at the upper right contained an atmospheric wind tunnel and the building immediately to its left was the world's first pressurized wind tunnel. The two temporary buildings in the upper left were engine dynamometer labs. The NACA also operated research aircraft and an airplane hangar on the flight line at Langley Field. (NASA L-01375)

letters "NACA" with eagle-like wings on either side was located directly above the main entrance when the laboratory opened in 1920. Similar winged logos were popular within many other military and civil organizations at the time. The symbol on the first NACA building was apparently the first use of a winged emblem by the NACA. The descriptive words "Research Laboratory" were engraved directly below the winged symbol. This layout appeared in original blueprints for the building as submitted by Deming for the approval of the NACA.[6] Our research did not reveal a formal directive by the NACA to Donn & Deming for the winged NACA symbol, and we conclude it is possible that the designers of the architectural firm were responsible for the first logo used by the NACA.[7]

The original NACA building underwent numerous renovations through the years, including side additions in the 1940s. In 1958, the new NASA organization continued to use the building for its Langley headquarters, but moved its administrative offices across Langley Field to the main NASA campus in the 1960s. NASA transferred the building to the U.S. Air Force in 1977. Today, the facility is known as the General Creech Conference Center. The original NACA logo, however, still remains embossed above the words "Research Laboratory."

6   Record Group 255, Box 78, Folder 21-1, NARA. Textual Reference Branch, College Park, MD.

7   In 1928, William I. Deming was asked to address a possible NACA desire to rename the building in honor of Charles D. Walcott. He examined the possibility of inscribing the words "Walcott Building" above the cornice of the doorway, but he suggested that the words "Research Laboratory" be retained as a designation of the type of building. His firm supplied a blueprint of the potential changes to the NACA. No further action took place on the renaming proposal. Record Group 255, Box 79, Folder 21-3, NARA Textual Reference Branch, College Park, MD.

FIGURE 1-2.

Entrance to the Langley Research Laboratory as it appeared in early 1928. The first example of a winged symbol used by the NACA was engraved directly above the building's name. (NASA L-02056)

FIGURE 1-3.

Close-up view of the NACA symbol. (NASA L-02056)

Emblems at the Flight Line

AIRCRAFT MARKINGS

In addition to the emblem exhibited on the NACA Research Laboratory building, many other logos were in evidence in the flight research organization. The symbols all appeared without evidence of formal directives and varied considerably during the early 1920s. The markings appeared on ground equipment and working clothes worn by mechanics during daily flight operations.

A small cadre of researchers at the LMAL under the direction of Edward P. Warner, Chief Physicist of the LMAL, began planning a series of studies to correlate wind-tunnel and in-flight results for aerodynamics even before the research laboratory was formally dedicated. In 1919, research flights began when the Army Air Service loaned two Curtiss JN-4H "Jenny" aircraft to the NACA for its embryonic studies. Although Warner's team added extensive instrumentation to the Jennies, the aircraft largely retained their original military markings with the exception of individual aircraft identification numbers. No special NACA logos were adopted at the time, setting a precedent that existed through most of the NACA years—that is, aircraft on temporary loan from the military carried no special NACA markings.

The first evidence of a winged emblem on an NACA aircraft occurred in 1923 on a U.S. Air Service de Havilland DH-9 biplane obtained by the NACA for use as a passenger airplane. The airplane, obtained on loan from the Army's McCook Field in Dayton, Ohio, arrived at Langley in 1922, and the rear cockpit was modified

FIGURE 1-4.

The NACA research airplane group in 1922 appears in standard military color schemes of the day with individual identification numbers added by the NACA. The aircraft on the far left is a Vought VE-7 and the others are Curtiss JN-4H Jennies. The second airplane from the right has been fitted with an auxiliary fuel tank on the upper center wing and tank-like sand containers on the lower wing for studies of roll response after the sand was released from one side. The Jenny on the right has been modified to use two identical wings with ailerons and a single cockpit enclosure for studies of improved performance. The aircraft were finished in yellow-gold linen with olive-green engine areas and vertical rudder stripes of (front-to-rear) blue, white, and red. (NASA L-00272)

by the NACA staff with an enclosed cabin in which two passengers could be accommodated in a face-to-face position.[8] The aircraft carried a winged emblem of interest forward of the cockpit on both sides of the fuselage. Unfortunately, the only two existing photographs of the airplane are of low resolution and do not permit the details of the emblem to be read. Examination of photographs of all other de Havilland aircraft flown at McCook Field during the period did not show such an emblem.[9] It appears, therefore, that

8   Eighth Annual Report of the National Advisory Committee for Aeronautics 1922 Including Technical Reports Nos. 133 to 158. 1 January 1923.

9   Telephone and e-mail correspondence with Brett A. Stolle, Manuscript Curator of the National Museum of the United States Air Force, Wright Patterson AFB, OH, 19 August 2014.

FIGURE 1-5.

A rather illegible winged emblem was displayed on this de Havilland DH-9 used for transportation by the NACA at Langley in 1923. Note the enclosed rear cockpit used by passengers. (NASA L-00503)

FIGURE 1-6.

This 1929 photograph of the Fairchild FC-2W2 owned by the NACA displays a winged symbol on the fuselage similar to that carried earlier by the de Havilland DH-9. Unfortunately, the words are unreadable. (NASA L-03546)

FIGURE 1-7.

The words identifying the FC-2W2 as a NACA aircraft are legible in this photograph taken during an entry in the Langley Propeller Research Tunnel in 1929. (NASA L-03778)

the emblem was applied to the DH-9 by NACA personnel after it arrived at Langley. The general arrangement of the emblem appears similar in some respects to what would become the NACA wings insignia a decade later.

The results of attempts to decipher the words of the emblem carried by the NACA DH-9 were very disappointing. The low-resolution photographs defied attempts at enhancement using current computer software and image enhancement methods. In addition, exhaustive searches within the archival NACA collections of NASA and NARA did not provide additional clarification beyond the operational usage of the airplane. Finally, since the aircraft was used solely for transporting passengers, no research reports on its use were written. Fortunately, further research on another NACA transport aircraft of the period provided a breakthrough in the mystery.

After three years of flying the DH-9 for transportation of personnel, the NACA replaced it by purchasing its first airplane, a more comfortable 4–6 place Fairchild FC-2W2 transport, in 1928. The transport initially appeared with what looks to be the same logo as that carried earlier by the DH-9—but once again

FIGURE 1-8.

Illustration of the NACA emblem carried on the FC-2W2 transport. (Contributed by Jennings Heilig)

only a few low-resolution flight-line photographs are available and the details of the words on the emblem are beyond recognition.

Fortunately, the Fairchild was used for research purposes as well as its transportation duties. The airplane was tested in the Langley Propeller Research Tunnel (PRT) in December 1929 for a project managed by Fred E. Weick to measure the aerodynamic drag of the airplane's components and the effectiveness of NACA engine cowl concepts to reduce drag and improve engine cooling.[10] Close-up photographs were taken during the tunnel entry. Nine of the ten photos taken were of poor quality, but one was of sufficiently high resolution to permit identification of the graphic elements and the words. The elements of the logo consisted of a shield with wings on either side. In the picture, the words on the emblem are (top to bottom) "National" "Advisory Committee" "for" "Aeronautics" and "Langley Field, VA."

Comparison of the general features of the logo with that of the DH-9 indicates that the earlier aircraft

probably carried the same marking. The DH-9 and the FC-2W2 were the only two NACA aircraft to have carried this insignia, and the emblem was removed from each vehicle during service in 1926 and 1930, respectively.[11] After these brief appearances, this particular logo was never used again by the NACA.

---

10 William H. Herrnstein, Jr.: Full-Scale Drag Tests on Various Parts of the Fairchild (FC-2W2) Cabin Monoplane. NACA TN 340, May 1930.

---

11 The DH-9 (serial number AS 31839) was referred to as Aircraft 9 at Langley and used for transportation of NACA personnel passengers (including Dr. George Lewis of Headquarters) and as a courtesy aircraft for the Air Service at Langley Field from 1923 to 1924. It was then completely modified by the NACA and re-entered service in 1926 with the designation DH-4M2 and was subsequently used for tests of a propeller dynamometer concept. After the conversion, it no longer carried a NACA emblem. See W.D. Grove and M.W. Green, "The Direct Measurement of Engine Power on an Airplane in Flight with a Hub Type Dynamometer," NACA Report No. 252, 1927. The FC-2W2 was used primarily for transportation of personnel between Langley and destinations such as the Washington, DC, Anacostia Naval Air Station, and the Naval Operations Center at Norfolk. It was also used for specific research studies including cowl tests, icing tests, and propeller brake testing. After the tests in the PRT were completed in late 1929, the airplane was repainted and the NACA emblem was removed. Daily logs of the usage of both airplanes are archived in the Langley Historic Archives.

THE SHIELD

In 1928, John F. Victory, Executive Secretary of the NACA, requested approval from the Secretary of Commerce for the design of a standardized symbol for use on the tail surfaces of the airplanes procured by the Committee.[12] The marking consisted of a shield with red and white vertical stripes beneath a blue header upon which the NACA number of the specific airplane was marked in silver. The designer of the shield emblem is unknown, but shield symbols were common for many organizations of the day and represented strength and unity. The proposal was approved by the Office of the Director for Aeronautics of the Department of Commerce and became the first systematic marking for NACA aircraft. The type of arrangement by which NACA aircraft were assigned—purchased, temporary, or permanent loan—dictated the markings applied.

GROUND CREWS AND EQUIPMENT

Winged insignias also appeared on ground support equipment and uniforms worn by the NACA flight research organization at Langley. Photographs of the 1920s show some commonality of logos for certain applications. Examples include a Ford truck modified with a Huck starter for NACA aircraft that carried a winged symbol with the letters "NACA" across the span of the emblem. The same logo adorned the back of coveralls worn by ground crew members. There is no evidence, however, that the emblem was common between aircraft and ground activities in the early 1920s.

The first NACA research aircraft were housed in a double-roofed hangar next to hangars of the Army Air Service adjacent to the main runways at Langley Field. These original hangars were identifiable by the

12 Letter from John F. Victory to the Secretary of Commerce dated 29 October 1928. See Arthur Pearcy, *Flying the Frontiers: NACA and NASA Experimental Aircraft*, Naval Institute Press, Annapolis, MD, 1993, p. 173.

FIGURE 1-9.

The standardized marking for NACA-owned aircraft appears on the vertical tail of a Fairchild XR2K-1 aircraft designated NACA 82 in this photo in 1940. The original black and white photograph on the bottom has been computer enhanced (colorized) to more vividly illustrate the color scheme and symbol carried by the silver and blue airplane. (Joseph Chambers based on NASA L-13581)

FIGURE 1-10.

A winged symbol denoting NACA equipment is displayed on this modified Ford with a Huck starter at Langley in 1926. The aircraft is a Vought VE-7 flown without a similar marking. (NASA EL-1997-00132)

FIGURE 1-11.

Aircraft mechanics wear coveralls with NACA symbols while modifying a Fokker Trimotor transport with a NACA cowling concept in 1929. (NASA L-03415)

letters "NACA" posted over the hangar entrance. In 1932, the flight research organization moved nearby to a new, larger building and the original hangars were demolished. As the result of a suggestion by Victory to Edward Sharp of LMAL in 1938, the roof of the new hangar was painted with the letters "NACA" in yellow on a 40-foot by 150-foot area of each side of the roof.[13] This identification scheme was later also applied to aircraft hangars at the NACA Ames Aeronautical Laboratory (AAL) at Moffett Field, California, and the NACA Aircraft Engine Research Laboratory (AERL) in Cleveland, Ohio, when the hangars were built during the 1940s.

---

13 Letter from Edward R. Sharp, Administrative Officer of LMAL, to NACA, 14 April 1938. Subject: Hangar Roof Aircraft Identification Sign. Record Group 255, Box 78, Folder 21-1, NARA Textual Reference Branch, College Park, MD.

FIGURE 1-12.

Early NACA markings in 1932. The aircraft, a Boeing PW-9, was the first American aircraft to have a metal fuselage and played a key role in a national program on the causes of in-flight structural failures. The NACA shield is visible on the rudder. Note the NACA letters above the entrance to the hangar. (NASA EL-1999-00433)

FIGURE 1-13.

Aerial view of Langley Field in 1939. The roof of the NACA hangar is readily identifiable by the large letters on its roof at the upper right of the photograph. (Army Air Corps)

FIGURE 1-14.

The roof of the Flight Research Laboratory at the NACA Ames Aeronautical Laboratory displays the NACA identifier in 1941. (NASA AAL-1449)

FIGURE 1-15.

The NACA Lewis laboratory flight hangar retained its NACA marking until the birth of NASA. (NASA C1955-38675)

Birth of the NACA Wings

In the early 1930s, virtually every federal agency had an emblem, including all of President Roosevelt's Depression-era programs. The technical, financial, and leadership scenarios of the NACA were changing, begging for the introduction of a new logo to represent the organization. The NACA needed to ask Congress for money since it was competing more directly with other services, and the workforce was expanding quickly and diversifying. Although research to date has not identified a formal document introducing the new symbol, the NACA Committee approved a new approach to the organizational logo. The basic layout was similar to the winged emblem carried earlier by the DH-9 and FC-2W2 aircraft, but the letters were shortened to NACA with periods after each letter. The new symbol would later evolve, become standardized, and known as the NACA wings insignia.

Applications of the NACA Wings

BUILDING EMBLEMS
Building of new NACA facilities at Langley rapidly progressed in the 1920s and 1930s. Following the introduction of the world's first pressurized wind tunnel in 1922, a series of revolutionary new tunnels

FIGURE 1-16.

The NACA emblem is seen above the entrance to the 8-Foot High-Speed Tunnel built in 1936. The tunnel circuit was demolished in 2011, and the office building is now operated by the Air Force. (NASA L-51079)

were conceived and put into operation. The 20-Foot Propeller Research Tunnel, the 30- by 60-Foot Full-Scale Tunnel, and the 8-Foot High-Speed Tunnel at Langley brought global attention and notoriety to the agency. Markings for new buildings followed the tradition used for the original laboratory—the NACA wings emblem was proudly positioned over the entranceway to the facilities, along with a name describing the mission of the organization housed in the building.

The expansion of the NACA in the late 1930s and early 1940s to include the AAL and the AERL further increased the number of NACA facilities displaying the NACA logo and building mission. Many of the buildings erected during the NACA years at the laboratories have been demolished, but several remain at NASA Centers today with their logos still on display.

FIGURE 1-17.

The Structures Research Laboratory building was the first building constructed in the NACA West Area in 1940, and it is still in use by NASA. The building displayed the early NACA wings insignia above the entrance doors to the right of the picture. (NASA L-22019)

FIGURE 1-18.

An early version of the NACA wings symbol as it appears today above the entrance door to the old NACA hangar at Langley. Note the description of the research focus under the symbol and the periods after each letter in NACA. The building was turned over to the Air Force in 1951, when the NACA built a huge hangar in the NASA West Area, but the exterior has changed very little. (Contributed by James Baldwin)

FIGURE 1-19.

The research staff of the Flight Research Laboratory at Ames in 1940 poses at the entrance to the laboratory. Note the subtle differences in the details of the NACA emblem compared to the details in the previous photograph. Each version of the early emblem was unique. (NASA AD88-0234-1)

FIGURE 1-20A.

The NACA emblem marks the location of the 16-Foot High-Speed Tunnel at Ames, which began operations in August 1941. The tunnel was converted to the 14-Foot Transonic Tunnel in 1953 and was demolished in 2006. The cleared grassy area is located at the center of Ames and is now referred to as NACA Park. The letters and wings symbol have a brass insert, while most other signs have the letters carved into stone. (Contributed by Glenn Bugos)

FIGURE 1-20B AND 1-20C.

The early NACA winged emblem appeared on the 7- by 10-Foot High-Speed Tunnel and the Technical Service Building at Ames. The Technical Service Building was one of the first operational buildings at Ames in 1940 and is still in existence today. (Contributed by Glenn Bugos)

FIGURE 1-21.

Test pilot Lawrence Clousing enters a Lockheed P-80 for a test flight at Ames in 1947. The Ames hangar was built in 1940. The NACA wing insignia is notable since it displays features of the early emblem but without periods after the letters. (NACA A-11880)

FIGURE 1-22.

FIGURE 1-23.

The 16-Foot Transonic Tunnel was put into operation at Langley in late 1941. Although the wind-tunnel circuit was demolished in 2011, the office building remains in use by NASA with the original NACA emblem above the entrance. (Contributed by James Baldwin)

The Langley Low-Turbulence Pressure Tunnel became operational in 1941. The photograph shows the original name for the facility—"Two-Dimensional Pressure Tunnel"—and the early NACA wing symbol. The tunnel and building were demolished in 2015 and the location is currently a grassy area. (Contributed by James Baldwin)

PUBLICATIONS AND CORRESPONDENCE

The NACA adhered to stringent directives for publications and external correspondence during the formative years covered by this era. For example, every reference to the NACA in technical reports was always made with periods after the individual letters. When spoken, individual letters were articulated by letter as N-A-C-A, rather than combined as the name "NASA" is usually pronounced today.

The NACA wings insignia appeared on a few documents as early as 1933. For example, one of the early Committee reports detailing the aeronautical challenges of the day and the facilities of the NACA at the LMAL carried the logo.

Interestingly, the logo did not appear on the covers of technical reports by the research staff. Covers of technical publications (e.g., technical reports and technical notes) carried only the author's name, the report publication number, and title of the publication without any insignia.

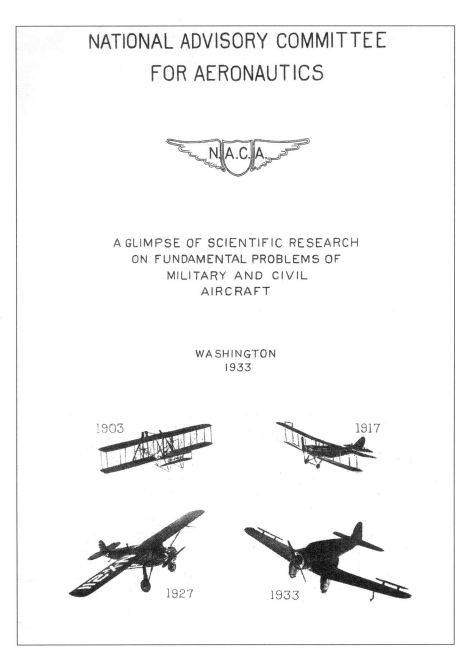

FIGURE 1-24.

The cover of an NACA overview document in 1933 displays the early NACA wings emblem. (U.S. National Archives at College Park, MD, Textual Reference Branch, RG 255, Box 275, Folder 53-3)

John Victory: No NACA Insignia

By the end of 1940, distinctly different versions of the NACA wings emblem were exhibited at many places throughout the laboratories and headquarters. Visitors might have easily concluded that the logo was the official insignia of the NACA, but that was not the case. Despite the growing usage of the wings, no official insignia for the NACA had been adopted.

In May 1940, Hardings Brothers, Inc., of Elmira, New York, sent a letter to the LMAL requesting permission to identify the National Advisory Committee for Aeronautics as one of the users of the company's products. (The specific product is not mentioned in the letter.)

Acting in his position as Executive Secretary of the NACA, Victory answered the request with these words:

> Your letter of May 22, addressed to our laboratory at Langley Field, Virginia, has been referred to this office for reply. There is no objection to your using the name of the National Advisory Committee for Aeronautics as one of the users of your products. The Committee does not have an official seal or insignia.[14]

Thus, the NACA had operated for its first quarter century without an official seal or emblem. The widespread use of the early NACA wings as an insignia had been an informal, decorative action that flourished within the organization—especially within the flight research community. That practice was to change in 1941 when a new version of the NACA wings was formally announced as the insignia of the agency. As will be discussed, a decade would pass before an official NACA seal would be designed and implemented.

14　Letter from John F. Victory to D. H. Laux, Vice President of Hardings Brothers, Inc., of Elmira, NY. 29 May 1940. Record Group 255, Box 275, Folder 53-3, NARA Textual Reference Branch, College Park, MD.

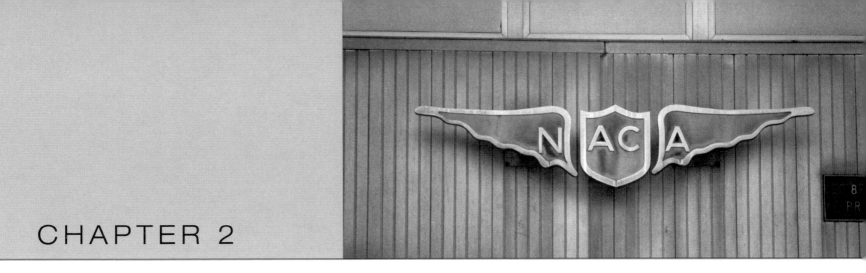

CHAPTER 2

Maturity and Pride, 1941–1957

Zenith of the NACA

The NACA continued its unprecedented acceleration of aeronautical research and technology development during World War II (WWII) and into the postwar years. Many regard the organization's contributions of that era as the most important accomplishments during its 43-year existence. The capabilities of the original NACA laboratory at Langley were enhanced by the addition of the AAL at Moffett Field, California; the Aeronautical Engine Research Laboratory in Cleveland, Ohio; the High-Speed Flight Station in Edwards, California; and Wallops Station in Wallops Island, Virginia. With the addition of unique facilities and technical expertise, an aura of team spirit and pride grew and prevailed among the laboratory staff members. During WWII, patriotic fever was rampant at all the laboratories, where staff typically worked six-day weeks with no vacations. Fundamental research was forced to take a backseat as the major focus of the NACA became support of military aviation—providing the facilities and expertise to contribute solutions to critical problems and conducting evaluations of emerging military aircraft and systems.

During the postwar years, the laboratories returned to areas of basic research, guided and challenged by lessons learned during the war and the emergence of radical new concepts—especially breakthrough technologies that would enable expanding the envelope into "higher, faster, and further" flight regimes. The NACA played a major role in the development and testing of a revolutionary series of X-planes that provided unprecedented data and information to the military and industry.

Although it had been unofficial, the early NACA wing insignia of the 1920s and 1930s was a common unifying symbol that was proudly displayed to represent the technical leadership of the agency. The symbols, however, were not consistent and varied for each application. The logo was standardized in 1941 by a formal NACA directive and became the official NACA insignia. It was modified in 1947 with a specific graphical change, but all versions retained the basic appearance of the earlier unofficial emblems. This chapter contains examples of applications of the standard and modified insignia to buildings, aircraft, personal items, technical reports, and correspondence during the era.

FIGURE 2-1A.

This photograph of blueprint LED-9535 shows the layout of the NACA standard insignia introduced in 1941 and a table of dimensional proportions. (National Archives and Records Administration, San Bruno, CA. Record Group 255.4.1. File G10-10)

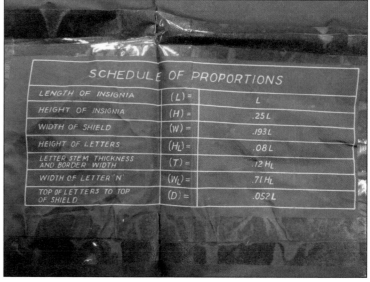

FIGURE 2-1B.

Close-up view of data providing schedule of proportions to be followed in the standard insignia. (National Archives and Records Administration, San Bruno, CA. Record Group 255.4.1. File G10-10)

An Official Insignia

The versions of the unofficial NACA wing emblem through 1940 had been very individualistic, with different details and a lack of uniform graphic content between applications. Although the components of the insignia consisted of a shield with separate wings and periods after each letter in NACA, the shapes and positions of the individual elements varied considerably.

On 24 April 1941, at the semiannual meeting of the Committee, a new official "NACA Standard Insignia" was approved for use on buildings under construction.[1] Details of the insignia were given in NACA blueprint LED-9535, created by the Langley graphics department. The new insignia again showed a shield, wings, and the letters "NACA" across the logo. The use of periods after the letters, however, was discontinued and specific proportions for details such as the length and width of the overall insignia; width of

---

1   Greatest Images of NASA (GRIN)-NASA Headquarters. *http://naca.larc.nasa.gov/search.jsp?R=GPN-2000-001539& qs=Ns%3DPublication-Date|0%26N%3D4294060006%2 B4294130770%26No%3D50* (accessed 13 August 2014).

FIGURE 2-2.

This view of notable NACA attendees at the NACA Executives Conference at Langley in February 1946 shows an unusual version of the NACA wings above the entrance of the 1941 addition to the Langley Administration Building. The logo appears to adhere to the proportions of the 1941 standard insignia directive but retains the periods after the letters per the early unofficial insignia. Attendees included (L–R): Front row: Smith De-France, AAL; Edward R. Sharp, AERL; John F. Victory, Edward H. Chamberlin, and John W. Crowley, Jr., NACA. Second row: H. J. E. Reid, LMAL; Carlton Kemper, AERL; Russell G. Robinson, NACA; Addison M. Rothrock, AERL; Ferril R. Nickle, AAL; Dolph Henry, LMAL. Back row: William J. McCann, AERL; Floyd L. Thompson, LMAL; Ralph E. Ulmer, NACA; W. Kemble Johnson, LMAL; Edward T. Mecutchen, NACA; Rufus O. House and Ernest Johnson, LMAL; Parmely C. Daniels, NACA; Elton W. Miller and T. Melvin Butler, LMAL; and Robert J. Lacklen, NACA. Lacklen would later become the Director of Personnel at NASA Headquarters and a key participant during the design and approval process for the NACA seal and the NASA seal and insignia. (LAL 47001)

the shield; the height, width, and thickness of the letters; and spacing between the top of the letters and the top of the shield were defined.

The directive of 1941 arrived at a time when the NACA was in the midst of building several major facilities at the AERL, and Langley and Ames laboratories, and instances arose in which the building markings did not adhere to the new order.

The "Modified Standard" Emblem

In 1947, a modified version of the official NACA insignia was introduced and applied; it was a simplified design of the standard insignia. This final version of the NACA insignia merged the separate elements of the shield and wings into a single solid background for enlarged NACA letters. Details of the emblem were given in NACA blueprint LED-16616, as supplied by the Langley graphics department. The design was widely used—particularly for NACA aircraft markings—and has been commonly remembered as the most famous NACA insignia.

During the planning for implementation of the modified standard insignia to Hangar Two at Ames in

1949, the NACA staff discovered that the proportions of the insignia shown in LED-16616 were in error.[2] The error was called to the attention of Headquarters and subsequently corrected. At the same time, Ames also requested permission to use the new version as a standard for NACA building entrance insignia. Headquarters denied the request.[3]

Victory, Executive Secretary of the NACA at the time, provided a detailed directive in 1952 that clarified the uses of the official and modified NACA wings insignia for the laboratories[4]:

1. On 24 April 1941, the Chairman, at the Semiannual Meeting, said he approved a Standard NACA Insignia for use on buildings under construction. The insignia showed a shield with the wing on each side inscribed with the letters "NACA" as presented on blue print LED-9535, but without periods.

2. In the years following, use of a Modified Standard NACA Insignia (LED-16616) for certain purposes has been approved. Appropriate uses of the Modified Standard NACA Insignia have been:

FIGURE 2-3.

The modified standard insignia of 1947 combined the shield and wings components of the earlier logos. (NACA LMAL 51669)

---

2   Letter from Arthur B. Freeman, Administrative Management Officer at Ames, to E.E. Miller, Chief of Division of Research Information, NACA Headquarters, 23 November 1949. National Archives and Records Administration, San Bruno, CA. Record Group 255.4.1. File G10-10 Procedure: Insignia-NASA Standard 1949–1952.

3   Letter from E.E. Miller, Chief of Division of Research Information, NACA Headquarters, to Ames, 9 December 1949. National Archives and Records Administration, San Bruno, CA. Record Group 255.4.1. File G10-10 Procedure: Insignia-NASA Standard 1949–1952.

4   Letter from John F. Victory to Langley, Ames, and Lewis, 22 May 1952. Record Group 255, Textual Records Branch, National Archives at College Park, MD. Also National Archives and Records Administration, San Bruno, CA. Record Group 255.4.1. File G10-10 Procedure: Insignia-NASA Standard 1949–1952.

(1) Report figures and tables

(2) Photographs

(3) Covers of RM's, TN's, and TM's

(4) Cover of telephone directories

(5) Slides

(6) Apprentice graduation invitations

(7) Inspection booklets

(8) Conference reports

(9) Movie titles

(10) Photostats of research data

(11) Laboratory directories

(12) Theses

(13) Speeches

(14) Laboratory forms

(15) NACA Civil Service Announcements

(16) NACA vehicles

(17) Decals

(18) Artist conceptions of new buildings

(19) Maintenance uniforms

(20) Basketball uniforms

(21) NACA research planes

3. Use of the Modified Standard NACA Insignia for purposes outlined in (2) is to be continued.

4. In the future, the Modified Standard NACA Insignia is to be used, instead of the Standard NACA Insignia, in such instances as on:

(1) Signal cover sheets

(2) Apprentice diplomas

(3) Testimonials

(4) Murals

(5) Displays for Air Force Day

(6) Displays for War Memorial Museum

(7) Apprentice graduation program

5. The NACA Standard Insignia will continue to be used on buildings.

Applications of Standard and Modified Wings

BUILDING EMBLEMS

The NACA insignia on new buildings and aircraft hangars at the NACA laboratories followed the directives of Headquarters after 1941. Each of the facilities exhibited impressively large versions of the insignia. The standard and modified wings insignia on the NACA flight hangars at the laboratories were particularly visible.

The NACA rapidly expanded its efforts in the construction of new wind tunnels in the years following

FIGURE 2-4.

The famous U.S. Army Air Forces B-17 bomber "Memphis Belle" visited the AERL on 7 July 1943 as part of its homecoming war bond tour after being the first bomber to complete 25 combat missions in Europe. The standard NACA wing insignia is displayed on the hangar. (NACA C1943-01866)

FIGURE 2-5.

A busy day of flight research at the NACA Aircraft Engine Research Laboratory in 1946. The AERL hangar was built in 1942 and displayed the standard NACA insignia. (NASA C1946-14736)

FIGURE 2-6.

The modified standard NACA wings logo appears on the Ames hangar (now NASA building N211) in this photo of a Republic F-84F during NACA tests in 1954. (GPN-2000-001509)

WWII. Replicas of the NACA wings were placed above the main entrance doors to the tunnel buildings. The Langley 8-Foot Transonic Pressure Tunnel was an important NACA facility used by Richard T. Whitcomb to develop several breakthrough concepts including winglets and supercritical-wing technology. The tunnel was constructed in 1953 and displayed the standard NACA insignia throughout its lifetime. The facility was closed in 1996 and demolished in 2011, but the tunnel's original NACA insignia is now displayed in the "America by Air" exhibit at the National Air and Space Museum (NASM) in the National Mall Building in Washington, DC. In honor of the NACA

centenary, NASA Administrator Charles Bolden formally presented the iconic logo sign to NASM Director Jack Dailey during a reception on 3 March 2015.

An important highlight of NACA activities in the postwar years was the creation of the High-Speed Flight Research Station at Muroc, California, for the X-1 project during the nation's assault on the sound barrier. That highly successful project was the first of countless achievements that have emanated from the complex at Edwards Air Force Base. The dedicated NACA staff initially occupied makeshift quarters that carried the NACA emblem at an area known as South Base. The original building housing the staff carried

FIGURE 2-7.

The NACA standard insignia was mounted above an entrance to the Langley 8-Foot Transonic Pressure Tunnel and remained there until the demolition of the building in 2011. The emblem is now on display at the National Air and Space Museum in the National Mall Building in Washington, DC. (Contributed by James Baldwin)

FIGURE 2-8.

The 12-Foot Pressure Tunnel, one of the most famous workhorse tunnels at the Ames Research Center, was identified by a facility sign with the NACA standard insignia. (Contributed by Glenn Bugos)

FIGURE 2-9.

The 40- by 80-Foot Tunnel at NASA Ames is the largest U.S. wind tunnel. It displays the NACA standard insignia. (Contributed by Glenn Bugos)

FIGURE 2-10.

The NACA staff at the High-Speed Flight Research Station poses in front of its South Base office building at Edwards, California, in 1950. At the time, the staff reported to NACA Langley. The modified standard wings emblem was on display. (NASA E33717)

the modified standard logo, but the modern building (Building 4800) that followed at Main Base exhibited the old standard NACA wings marking in compliance with the 1941 directive.

Use of the NACA insignia on new buildings was reserved for major facilities and did not, in general, apply to secondary buildings.

AIRCRAFT MARKINGS

From 1946 to 1957, the NACA's expanding fleet of research aircraft went through dramatic changes in terms of color and insignia. In 1946, the NACA, the United States Air Force (USAF), and Bell Aircraft pursued the development of a radical series of experimental research aircraft, which became known as the Bell X-1 series—specifically designed to investigate and conquer the mysterious sound barrier. The three early X-1s

FIGURE 2-11.

By 1954, the installation was renamed the NACA High-Speed Station and became an autonomous unit reporting to NACA Headquarters. Interestingly, the newly constructed main building at Main Base exhibited the older standard logo. (NASA E-33718)

FIGURE 2-12.

The entrance to the Ames Aeronautical Laboratory exhibited the NACA modified standard wings emblem. (NASA A-13425)

FIGURE 2-13.

Not all NACA buildings displayed the NACA insignia. For example, the security guard house at the Lewis laboratory used individual letters in 1956. (NASA GRC 1956_43269)

FIGURE 2-14.

FIGURE 2-15.

The second X-1 research airplane was used by the NACA for specific flights of interest to the research community. The airplane was originally orange in color but was painted all white and displayed the old NACA shield on the vertical tail in 1949. (NASA E49-009)

Two years later, in 1951, X-1-2 retained its white paint finish, but the NACA shield was replaced with the modified standard NACA wings insignia in a gold tail band. The X symbols were painted on the fuselage for reference during visual measurements of aircraft attitude. (NASA E52-0670)

were initially painted bright orange for presumed ease of visibility and tracking during flight. During Chuck Yeager's historic X-1 flight in October 1947, however, the aircraft was difficult to see. Flight research for the second X-1 was managed by the NACA for its specific research interests and was painted a more easily discernible white. X-1-2 initially displayed the NACA shield to honor the traditional shield markings from the pioneering days of the agency. After its first flights, the vertical tail surface was refinished with the modified standard NACA wings insignia within a gold tail band. Archival searches did not identify a reference for the foregoing applications.

Other experimental NACA aircraft of the 1950s adopted the highly visible white paint scheme with a markings progression similar to that of the X-1 program. For example, the paint scheme for the Douglas D-558-1 Skystreak was changed from scarlet to white, and its tail emblem transitioned from the NACA shield to the gold NACA wings logo. Other research aircraft adopting the overall white scheme included the swept-wing Douglas D-558-2, designed to further explore the realm of supersonic flight beyond Mach 1; the Douglas X-3 Stilleto, designed to advance the efficiency of supersonic flight; the Northrop X-4, designed to evaluate the characteristics of tailless jet

FIGURE 2-16.

The Douglas D-558-1 shown in its early markings with the NACA shield on the vertical tail (top) and in flight with the NACA wings band (bottom) in the early 1950s.

FIGURE 2-17.

The swept-wing NACA/Navy Douglas D-558-2 displays the NACA shield in 1949 (top) and the NACA wings band in 1955 (bottom). (NASA E49-00200 and NASA E-1442)

FIGURE 2-18.

White NACA X-planes on display at the High-Speed Flight Research Station. Aircraft are (left to right): D-558-2, D-558-1, X-5, X-1, XF-92A, and X-4. Photo of NACA research aircraft in front of the South Base hangar was taken on 1 March 1952. Both the aircraft and hangar exhibit the modified standard wings logo. (NASA EC-145)

FIGURE 2-19.

This Grumman F9F-2 was flown at Langley in 1954 to evaluate a pioneering NACA analog fly-by-wire concept. In addition to the gold NACA tail band, it carries NACA letters on the rear fuselage. (EL-2002-00295)

FIGURE 2-20.

In some cases, research aircraft carried the NACA insignia on the forward fuselage as shown here. Ames Director Smith DeFrance poses with the last Chairman of the NACA, James H. Doolittle, in the Ames hangar with a Convair F-102 airplane. (A-23459)

FIGURE 2-21.

The gold NACA tail band is carried by this B-57 aircraft at Lewis in 1957. (NASA 1957_45904)

FIGURE 2-22.

Striking photograph of an F-100 with NACA and U.S. Air Force markings during a flight-test program on inertial coupling phenomena at the NACA High-Speed Flight Station in 1957. (E-3213)

aircraft; and the Bell X-5, designed to study the variable-sweep wing concept.

From the mid-1950s through 1957, virtually all NACA research aircraft at the laboratories followed the lead of the Muroc group and carried the modified standard NACA wings enclosed within a gold tail band. The following photographs illustrate markings for a few of the aircraft of the era.

Flight research at the NACA laboratories was extremely healthy and productive during the 1950s. Requests from the military for evaluations of advanced aircraft were abundant, and basic in-flight research studies included correlation with data from ground-based experiments and wind tunnels.

FIGURE 2-23.

A McDonnell F-101 was flown at Langley in 1956 for investigations of sonic boom phenomena. The project produced data used today in the United States sonic boom data-base. The airplane carries a large NACA tail band as well as Air Force mark-ings. (EL-2002-00289)

FIGURE 2-24.

In 1957, Langley con-ducted flight tests of a Grumman F-11F-1 to obtain data for correlation with wind-tunnel and the-oretical results obtained on NACA aerodynamicist Richard Whitcomb's area rule. The NACA tail band lacked the gold color typically carried by other research aircraft. (NACA LAL 57-2253)

FIGURE 2-25.

Ames test pilot George Cooper poses with an F-100 that displays the modified standard wings insignia under the forward fuselage. (NASA A-22548)

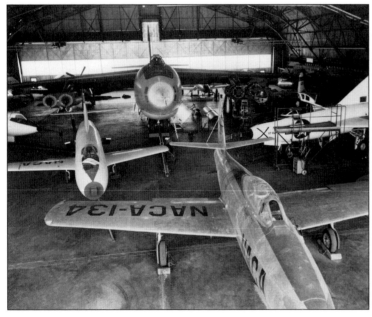

FIGURE 2-26.

In the 1950s, NACA flight hangars were busy with numerous high-priority flight projects. The Langley hangar (top) housed three generations of aircraft wearing the NACA tail band, including (bottom to top) a P-51H, an F-101A, and an F-86A. The High-Speed Flight Station hangar (bottom) housed aircraft with a variety of NACA markings including tail bands and wing identification numbers. (NASA LAL 96951 and NASA E-959)

INSIGNIA AND DECEPTION

One of the most remarkable activities in the history of the NACA involved the use of the NACA insignia for military deception during the late 1950s period of the Cold War. At that time, high-altitude overflights were being conducted for espionage and spying by the United States. During the startup of U-2 spy plane missions over Europe and the Soviet Union in the mid-1950s, the Central Intelligence Agency (CIA) and other agencies required a cover story for the covert missions. After extended discussions and consideration by officials at the highest levels, a cover story was created that the U-2 was being used for high-altitude weather research by the NACA.[5]

Dr. Hugh L. Dryden, Director of the NACA, gave his approval for the cover story, and a NACA press release on 7 May 1956 stated:

> The Lockheed U-2 shown here is being used by the National Advisory Committee for Aeronautics, to obtain detailed information about gust-meteorological conditions at high-altitude. The research program makes use of instrumentation furnished by the NACA and the Wright Air Development Center of the USAF and logistical and technical support is provided by the Air Weather Service of the USAF. Since the program began last spring, numerous data gathering flights have been made in the United States and elsewhere in the world. The NACA has found the U-2 (Powered by a Pratt & Whitney J-57) a most useful research tool, especially because of its ability to maintain flight at high-altitude, as high as 55,000 feet. Subjects under study include clear-air turbulence, convective clouds, and the jet stream.[6]

In reality, the NACA never used the aircraft, although NACA instrumentation was carried on flights. Pilot training of CIA pilots had begun in 1955 at Groom Lake, Nevada (also known as "The Ranch"), using similarly marked silver U-2 aircraft. Each aircraft had

---

5   For an in-depth discussion of the NACA U-2 activities, see Peter W. Merlin, *Unlimited Horizons: Design and Development of the U-2* (NASA SP-2014-620).

6   Photo and photo caption in NASA file folder, Record Group 255, Still Pictures Branch, National Archives at College Park, MD.

a fictitious NACA tail band and number on the tail and the letters "NACA" painted on the lower left wing. The NACA cover story was maintained until 1 May 1960, when CIA pilot Francis Gary Powers was shot down while flying over Soviet Union airspace.[7]

After Powers was shot down, an all-black U-2 was hastily painted with a fictitious tail number and NASA markings (by then the NACA had been absorbed by NASA) and put on view for the press at the NASA Flight Research Center with a cover story about a U-2 conducting weather research that may have strayed off course after the pilot reported difficulties with his oxygen equipment. In 1971, NASA finally took possession of two U-2s for high-altitude research.

Employee Applications

LAPEL PINS

The employees of the NACA fostered a culture of pride and dedication that extended to an appreciation and respect for the organization's insignia as embodied in decorative pins, service awards, and awards. Display of the insignia on personal and government items was a common habit. In 1943, after much consideration of a lapel pin for NACA personnel and a review of the designs submitted, Victory announced that the standard NACA insignia had been selected for the pin.[8]

---

7   For discussions of the U-2 story, see Norman Polmar, *Spyplane: The U-2 History Declassified,* Zenith Press, 2001; Dennis R. Jenkins and Tony R. Landis, "Date with the Dragon Lady. The Amazing Story of the Versatile Lockheed U-2," *Wings* magazine, August 2005, pp. 28–61; Zaur Eylanbekov, "Project Aquatone," *Air Force* magazine, July 2010. pp. 51–57; and Jennifer E. Davis, "There is No Good Time for Failure: Examining NASA's Involvement in the U-2 Spy Plane Cover Story," Master's Thesis, Utah State University Department of History, 2005.

8   *LMAL Bulletin,* Langley internal newsletter, 5 June 1943, p. 3.

FIGURE 2-27.

This NACA-marked version of the Lockheed U-2 spy plane was one of about three dozen that participated in training of pilots at Groom Lake, Nevada, by the Central Intelligence Agency (CIA). NACA Director Hugh Dryden agreed to a cover story for the CIA to protect the true intentions of the aircraft missions. (NACA LAL 57–96 via NARA at College Park, MD, RG 255-RF series, Langley Aeronautical Laboratory, Flight Research)

FIGURE 2-28.

An all-black U-2 was put on display at the NASA Flight Research Center in 1960 after the F. Gary Powers incident. (NASA GPN-2000-000112)

FIGURE 2-29.

Lapel pin offered for purchase by NACA employees beginning in 1943. (Contributed by Peter W. Merlin)

The pin, suitable for wear by both men and women, had an overall length of about three-quarters of an inch and could be purchased by employees at the NACA laboratories and headquarters. The NACA insignia design was finished in red, white, and blue with the letters "US" and "NACA" in silver.[9]

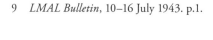

9   *LMAL Bulletin*, 10–16 July 1943. p.1.

FIGURE 2-30.

Icing expert Lewis Rodert of the Ames laboratory wears the red, white, and blue NACA pin while accepting the Collier Trophy from President Truman in 1947. Rodert had started his career in icing at Langley, then transferred to Ames, and was employed at the AERL at the time of the award. (NASA C47_20298)

FIGURE 2-31.

NACA emblems abound in this photo taken in 1957 at Ames during a visit from celebrity Arthur M. Godfrey. Godfrey was a certified pilot and had a strong interest in aviation. The group includes (left to right) Godfrey, Ames test pilot George Cooper, and Ames Director Smith DeFrance. Note the NACA insignia on the vehicle in the background and on the patch on Cooper's shoulder. DeFrance is wearing the NACA lapel pin. (NASA A-23317)

FIGURE 2-32.

These designs for the NACA meritorious service pins were submitted by J. J. Lankes of Langley (top) and Harry DeVoto of Ames (bottom). (Langley *Air Scoop*, internal newsletter, 4 March 1949, Volume 9, Issue 8, p. 3)

SERVICE PINS

An administrative item of great interest at the NACA sites occurred in 1948, when it was decided to award meritorious service emblems to members of the NACA staff upon completion of 20 years or more service. All of the NACA laboratories were invited to propose designs for the pin emblem that would be "symbolic of the NACA and which will also be striking in appearance."[10] Laboratory employees at Langley and Ames contributed both serious and humorous designs to NACA Headquarters for selection of the final design.

Two of the most talented artists employed by the NACA responded with several ideas for designs. Julius J. Lankes, head of illustrators at Langley, was the creator of many impressive works.[11] Noted artist Harry DeVoto of Ames also contributed several potential candidates for the service pins. DeVoto transferred to Ames from Langley where he had been the former art editor of the *Air Scoop* newsletter. Lankes would leave the NACA in 1950, but DeVoto remained at Ames through the establishment of NASA. DeVoto led the Ames proposals for the design of the NASA insignia and seal in 1958, as will be discussed in a later chapter.[12]

One of the notable designs for the service pins was submitted by F. D. Morris of the East Photo Lab at Langley. Morris proposed the use of the modified standard NACA wings with the years of service indicated by number on an auxiliary tab of some sort. His submittal apparently had a large influence on the ultimate selection.

In early 1949, a survey of the NACA staff was conducted to determine the most desirable design submission. The service pin design selected by laboratory directors and NACA Headquarters resembled the modified standard NACA wings emblem, but with the numbers 15, 20, 25, or 30 added to represent years of service—an arrangement very similar to the scheme submitted by F. D. Morris. All pins would be gold; the 20-year pin would have a ruby stone, the 25-year pin an emerald, and the 30-year pin a diamond.[13]

10 Langley *Air Scoop*, in-house newsletter, 28 October 1948, Volume 7, Issue 43. p. 4.

11 See *http://crgis.ndc.nasa.gov/historic/NACA_Memorial_Aeronautical_Laboratory#Aeronautical _History_Murals* (accessed 29 October 2014), *http://history.nasa.gov/nltr31-1.pdf* (accessed 29 October 2014), and *http://history.nasa.gov/nltr31-2.pdf* (accessed 29 October 2014).

12 "Ames Research Center View of NASA Logo Design," Letter from Harry J. DeVoto to Steve Garber, 6 May 2001, Historical Reference Collection (HRC), file 61.

13 Langley *Air Scoop*, 4 March 1949, Volume 9, Issue 8 p. 3.

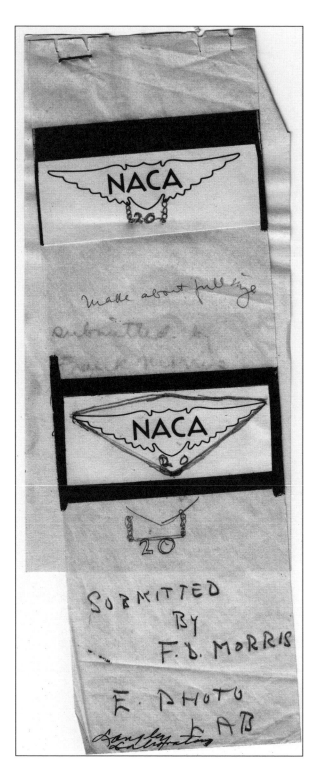

FIGURE 2-33.

Sketch of the 20-year service pin submitted by F. D. Morris of Langley. (Record Group 255, Box 275, Folder 53-3, NARA Textual Reference Branch, College Park, MD)

FIGURE 2-34.

NACA 15-year service pin awarded to Ben A. Goddin of Langley. Note the similarity to the sketches by F. D. Morris in the previous figure. (Contributed by Andy Goddin)

FIGURE 2-35.

Dr. Hugh L. Dryden held the position of Director of the NACA from 1947 until October 1958. He was Deputy Administrator of NASA until his death in December 1965. In this photo, Dr. Dryden is wearing a 10-year NACA service pin. (NASA GPN-2002-000105)

BADGES

Security at the NACA laboratories and headquarters during WWII was a major issue as concerns over sabotage and espionage intensified with the recognition that spies and informants were active. The situation was especially severe at Langley, which was adjacent to numerous Army and Navy installations in an area known to be a hotbed of spying activity. Identification badges for personnel and their vehicles were instituted. Staff members were required to display personal badges with their photographs, and they were required to display metal tags on automobiles for entry to the laboratory. Use of badges was strictly enforced by security guards.

Badges and other identification cards were also displayed by Committee members of the NACA for entry to the NACA laboratories and other government facilities. Keith Yoerg, great-grandnephew of Orville Wright, has inherited Wright's badge for access to the U.S. Army Air Forces Materiel Center at Wright Field in Dayton, Ohio.[14] The badge identifies his affiliation with the NACA and honors him with the first access badge to his namesake laboratory.

---

14 "A Legacy of Flight," *The Researcher News,* Langley's in-house newsletter, *http://researchernews.larc.nasa.gov/archives/2005/72905/legacy.html* (accessed 19 August 2014).

FIGURE 2-36.

Orville Wright's badge for access to the U.S. Army Air Forces Materiel Center at Wright Field. The background for the badge in the photograph is original fabric from the 1903 Wright Flyer. (Contributed by Keith Yoerg)

FIGURE 2-37.

Other identification cards were carried by members of the NACA Committee. This card identifies Edward V. "Eddie" Rickenbacker as a member of the Committee from 1956 to 1958. (Contributed by Peter V. Merlin)

FIGURE 2-38.

The NACA standard wings were a prominent component of the NACA identification badge worn by O. W. Culpepper of Langley in 1942. The bluish-purple color indicates an unclassified security clearance. (NASA Langley LAH)

FIGURE 2-39.

Clarence J. Parker of Langley wore an orange badge, indicating that he was cleared at the Confidential level. (Contributed by Tracey Redman)

Employees received frequent reminders about the placement and importance of personnel badges to ensure that the badges were worn on the upper portion of the employees' clothing where they would be visible at all times. (Belt, wallet, and purse locations were not satisfactory.) An article in the *LMAL Bulletin*, Langley's in-house newspaper, stated:

> It has been mentioned that women are the chief violators of this regulation, many of them claiming that the pin part of the badge makes holes in their clothing. The exercising of a little ingenuity will help eliminate this problem, officials declared.[15]

The ingenuity exhibited by employees included using small metal clamps to grasp the offending pin on the backside of the badge and using a necklace-type band to hold the badge around the neck. The security clearance level of the individual was indicated by the color of the badge, with bluish-purple denoting an unclassified level, orange indicating Confidential, and red signifying Secret.[16]

In 1952, the security organization at Langley made a major change away from the hole-producing pin-type badges. Instead, flat plastic clip-on badges were introduced that would not perforate the wearer's apparel.[17]

---

15   *LMAL Bulletin*, Issue 19, Volume 2, 28 August–3 September 1943, p. 5.

16   P. Kenneth Pierpont, Langley retiree, interview by Joseph Chambers, Newport News, VA, 23 August 2014.

17   Langley *Air Scoop* newsletter, Volume 11 Issue 45, 7 November 1952, p. 4.

FIGURE 2-40A AND 2-40B.

Ames Director Smith DeFrance (left) was presented with his 35-year service award by his assistant, Jack Parsons (right). Parson's NACA badge is shown in the enlargement. (NASA GPN-2000-001525)

FIGURE 2-41.

Ames test pilot George Cooper (left) and Ames Director Smith DeFrance (right) pose next to an F-86 research aircraft. DeFrance is wearing his NACA badge. (NASA A-16539)

FIGURE 2-42.

Frank Bechtel began his 33-year NACA/NASA career at the NACA Aircraft Engine Research Laboratory in 1941. Here, he displays his AERL badge from the early 1940s. The red badge denotes a Secret clearance. (Contributed by Scott Marabito)

FIGURE 2-43.

Langley's internal newsletter, *Air Scoop,* illustrated the new clip-on badges in November 1952. Samples included permanent and temporary badges. (NASA Langley Historical Archives)

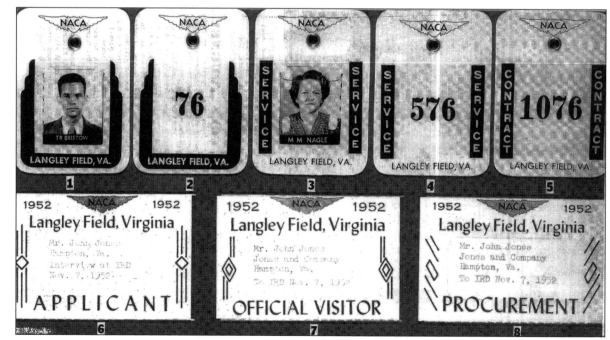

FIGURE 2-44.

Honorary NACA plaque presented to Fred E. Weick in recognition of his contributions to aviation. The plaque emblem is the standard design commonly used on retirement plaques. (NASA Langley Historical Archives, Weick collection)

RETIREMENT PLAQUES

Perhaps the most revered award bestowed upon retirees who worked for the NACA was a plaque adorned with the standard NACA wing insignia. NASA retirees who did not work for the NACA during their career are not eligible to receive this coveted plaque. The practice—extended well into the NASA years—was greatly appreciated by the dedicated employees whose hard work and team spirit contributed so much to the advancement of aeronautics and international respect for the agency.

Employees at all levels received such plaques—from administrative personnel, to top-ranking researchers and managers, to administrative personnel and support technicians. At a special event during a visit to Langley in 1990, an honorary NACA plaque was presented to Fred E. Weick, a famous researcher who led critical studies in the Atmospheric Wind Tunnel (AWT), the Propeller Research Tunnel (PRT), and

airplane flight tests.[18] His work in the PRT resulted in the NACA receiving its first Collier Trophy for the NACA cowling concept.

PUBLICATIONS AND CORRESPONDENCE

Technical publications produced by the staffs of the NACA laboratories did not feature emblems until about 1944. Prior to that year, NACA reports were published with a cover that simply stated the title, the author(s), and the date of publication. From 1944 to about 1949, covers of lower-level publications known as the Technical Note series carried the standard NACA wings. After 1949, the modified standard wings were used.

18 Weick came to the NACA in 1925 and left in 1936. He did not retire from the NACA.

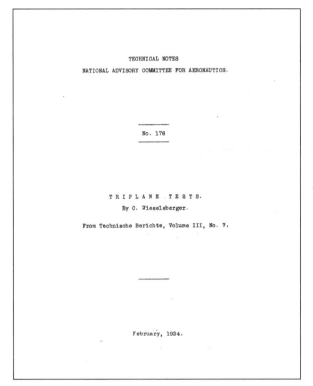

FIGURE 2-45.

Covers of early NACA technical reports did not feature symbols or insignia until about 1944. As shown here, many of the publications were translations of research conducted by other organizations, including universities and European research laboratories.

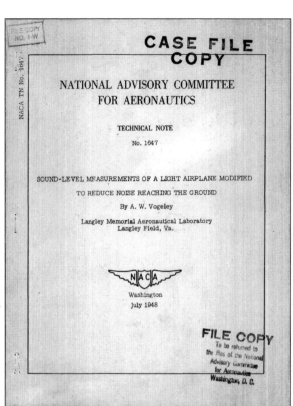

FIGURE 2-46.

The NACA standard insignia was used on the cover of NACA reports known as technical notes from 1944 to 1949.

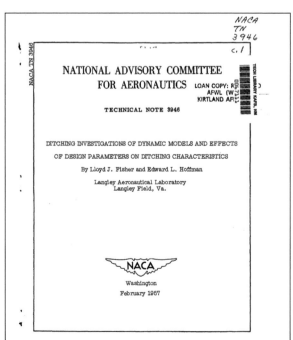

FIGURE 2-47.

After the modified standard insignia was adopted, it was used on the cover of NACA technical note publications from 1949 until the end of the NACA in 1958.

MISCELLANEOUS APPLICATIONS

The unifying symbolism of the NACA wings spread to other areas of daily life at the laboratories. Sports uniforms worn by enthusiastic participants from various organizations in intense yearlong in-house competition often carried the wings insignia, as did NACA apprentice school emblems and activities. The apprentice school was a proud element of the NACA tradition that was extremely productive in educating masses of future NACA employees.[19]

Pride in the NACA insignia spilled over into private lives as well. Personal items, such as pencils and other memorabilia, could be purchased by employees; and some items, such as drinking glasses and mugs, can still be purchased today.

Former NACA employees have cherished the memories of their years at the laboratories and headquarters long after their retirement. Special events, such as reunions, have been notable events in many lives. The first reunion took place in 1976 at Asheville, North Carolina, and subsequent reunions took place on a more or less regular basis beginning in 1982. Over 360 attended the 12th and final reunion of NACA employees in May 2008 at the NASA Langley Research Center. The 1988 reunion held in San Jose, California, had a peak attendance with 810 people.

Other souvenirs of the NACA years include special tokens and medallions in remembrance of events and the significant contributions of the agency. One special medallion minted for the employees of the NACA contained material from a Curtiss JN-4 "Jenny" tested during the first aircraft flight research program at the LMAL and material from the Bell X-1 series aircraft

FIGURE 2-48.

NACA Apprentice School banner bears the NACA modified wing emblem. Date unknown. (Contributed by James V. Plant)

---

19 "Apprentice to Expert: Langley Grows Its Own," available online at *http://articles.dailypress.com/1990-08-20/news/9008200098_1_apprentice-program-nasa-langley-research-center-blades* (accessed 6 September 2014). The Apprentice School at Langley was set up in 1941 by the Apprentice Administration in conjunction with officials of the War Training Program. Under the NACA Apprenticeship Standards approved by the Federal Committee on Apprenticeship, students were required to obtain 150 hours of study each year to become machinists, aircraft sheet metal workers, model makers, and wind-tunnel mechanics. Members of the laboratory staff, including both engineers and tradesmen, served as classroom instructors. The first class of Langley apprentices, made up of 14 students, graduated 17 February 1943. *LMAL Bulletin,* in-house newsletter, Volume 2 Issue 6, 5 April 1943.

flown at the NACA High Speed Flight Station. The dedication statement read:

> The contributions by NACA to the history of aviation during the span of time between these two aircraft will be felt by all aircraft to come and best remembered by those who were involved in these and other aeronautical activities during such an important part of our Nation's technological heritage.

FIGURE 2-49.

This NACA reunion brochure from the second reunion at Williamsburg, VA, in 1982 exhibited the beloved wings and seal. (Fred Weick Collection LHA)

FIGURE 2-50.

A special award of NACA wings was presented to James J. Modarelli, who served as Chairman of the NACA Reunion III held at the NASA Lewis Research Center in 1985. Modarelli was the creator of the NASA seal and insignia. (Contributed by James J. Modarelli, Jr.)

FIGURE 2-51.

Special medallion containing materials from a Curtiss Jenny biplane and one of the Bell X-1 research aircraft. (NASA Langley LHA)

The NACA Seal

The NACA wings insignia has often been mistakenly referred to as the NACA seal, but an actual seal emblem was not developed and approved until the 1950s. The fact that the NACA did not have a seal until then was a surprising revelation—particularly since seals were common in most government organizations. Insignia are almost informal logos adapted for general representation of organizations, whereas seals are usually formal symbols used by high-level officials—typically for more restricted uses. At a meeting of the Executive Committee on 23 January 1953, a proposed design of an official seal for the NACA was submitted for approval on the basis of a rather obscure requirement.

The seal was stated to be necessary "For use on the certificates issued to graduates of the apprentice training courses conducted at the NACA laboratories, and also for use on the commissions of appointment to NACA technical committees and subcommittees."[20] An initial design proposal had included the NACA insignia, which was eliminated by agreement of the attendees.

On 4 February, the Committee requested that the Heraldic Branch of the Office of the Quartermaster General of the Department of the Army prepare an official seal based on a drawing that the NACA had informally discussed and left with the branch. The request was for a plaque, line drawings, a colored painting, and a proposed executive order approving the seal to be presented by the NACA to the Bureau of the Budget for approval by the President.[21] On August 12, the Office of the Quartermaster General responded to the NACA with all the requested deliverables. The Committee then sent all materials to the Bureau of the Budget for transmittal to the President of the United States for approval.[22]

The design for the official seal showed a drawing of the takeoff of the Wright brothers' airplane at Kitty Hawk, North Carolina, on 17 December 1903, with a circular border displaying the words "NATIONAL ADVISORY COMMITTEE FOR AERONAUTICS U.S.A."

20 Minutes of Executive Committee meeting, 23 January 1953, p. 5, Langley Historical Archives, Milton Ames Collection.

21 Letter from John F. Victory to the Quartermaster General, Department of the Army. 4 February 1953. Files of the Institute of Heraldry, Department of the Army, Fort Belvoir, VA, to be deposited in the NASA HRC.

22 Letter from William D. Jackson, Chief, Research and Development Division, to Robert Lacklen, NACA Headquarters, 12 August 1953. Files of the Institute of Heraldry, Department of the Army, Fort Belvoir, VA, to be deposited in the NASA HRC.

the lower wing and Wilbur Wright running alongside, just after letting go of the right wing tip.

3. The Government Printing Office is preparing cuts of the seal for use on NACA publications and possibly on letterheads. An official impression seal is being prepared by the Bureau of Engraving and Printing.

4. The official seal in color was prepared by the Heraldic Branch, Office of the Quartermaster General, Department of the Army, and is mounted on the wall in the Board Room at Headquarters.

The NACA laboratories were advised of the formal adoption of the new NACA seal in a letter from John F. Victory[23]:

1. On 14 October 1953, President Eisenhower issued Executive Order 10492, approving an official seal for the NACA.

2. The seal is a representation of the famous photograph depicting the first successful flight of an airplane. The seal shows the first flight of the Wright brothers' Kitty Hawk Flyer taking off from the launching rail on the sandy soil at the foot of Kill Devil Hill, NC, with Orville Wright lying prone on

As will be discussed in the next chapter, many of the individuals involved in the establishment of the NACA seal also had roles in the development and establishment of the NASA insignia and seal a few years later.

The NACA seal was used in the design of special awards presented to employees. Among these were the NACA Distinguished Service Medal and the NACA Exceptional Service Medal.

The seal was displayed in a prominent location in the boardroom at the NACA Headquarters building in Washington, DC. Visitors to the boardroom at NACA Headquarters were also impressed by a detailed mural depicting the location and scope of the NACA facilities across the United States. The mural had been created in 1950 by Langley illustrators to signify the breadth of the Committee's facilities.

---

23 Letter from John F. Victory to NACA Headquarters, Langley, Ames, Lewis, NACA High-Speed Flight Research Station, NACA Western Coordination Office, NACA Liaison Office at Wright-Patterson Air Force Base, 28 October 1953, Langley Historical Archives, Milton Ames Collection.

FIGURE 2-53.

The NACA Distinguished Service Medal (left) and the NACA Exceptional Service Medal (right). (NASA Ames Artifacts Collection ART1387348 and ART1387349)

FIGURE 2-54.

The NACA Exceptional Service Medal is presented at the NACA High-Speed Flight Station in November 1956 by Dr. Hugh Dryden to the crew of the X-1A research aircraft. L–R: Dr. Dryden, Joe Walker (X-1A research pilot), Stan Butchart (pilot of the B-29 mothership), and Richard Payne (X-1A crew chief). (NASA E-2672)

FIGURE 2-55.

This mural featuring the standard NACA wings and sketches of the NACA facilities was displayed in the NACA Headquarters boardroom along with the NACA seal. (LAL-67805)

Anticipating the Future

Working at the leading edge of technologies in aeronautics, the NACA laboratories joined in the national push toward higher, faster flight capabilities. Interest in hypersonic flight (speeds above Mach 5) had been stimulated as early as 1945 in part because of German applications of V-2 missiles and advanced ground facilities near the end of WWII.[24] In 1952, NACA laboratories began studying numerous problems likely to be encountered in flights to space, and in May 1954, the NACA proposed to the Air Force development of a piloted research vehicle that would study the problems of flight in the upper atmosphere and at hypersonic speeds. That vehicle would become the famed rocket-propelled X-15 hypersonic research airplane.

As efforts to achieve spaceflight intensified, other nations made great strides toward achieving the same goal, and, on 4 October 1957, the Soviet Union launched a satellite known as Sputnik, which led to a massive effort by the United States to gain the lead in spaceflight. One result of this effort was the creation of a new government agency that would absorb the personnel, facilities, and "can do" pride of the NACA. Along with the new Agency would come a new insignia and a new seal.

24 James R. Hansen, *Engineer in Charge: A History of the Langley Aeronautical Laboratory, 1917–1958.* (NASA SP-4305, 1987).

CHAPTER 3

A New Agency, 1958–1959

Search for a New Logo

Following the launch of Sputnik, the wave of national enthusiasm in support of the nation's aeronautical progress resulted in a new agency with a logo to inspire efforts to meet the space race challenge. Today, the official NASA seal and the less-formal NASA meatball insignia are among the most-recognized emblems throughout the world. Less known, however, is the story of how the emblems were conceived. The logos, which include symbols representing the space and aeronautics missions of NASA, were implemented in 1959 and were influenced by activities at three NASA Centers: the Lewis Research Center in Cleveland, Ohio; the Ames Research Center at Moffett Field, California; and the Langley Research Center in Hampton, Virginia. A particularly interesting element of the story is how the "red wing" graphic representing aeronautics in the emblems was originated and how its shape was inspired by actual research programs. Unfortunately, a great deal of confusion and misinformation now exists in the literature regarding the wing

emblem, which has erroneously been referred to as a "slash," "vector," "airfoil," "hypersonic wing design," and even as an "alternate shape of the constellation Andromeda." Although some aspects of the origins of the NASA seal and insignia have been documented in the past, this chapter collates and summarizes the details of the birth of the NASA logos, including previously unpublished information and a discussion of the roles of other government agencies in the creation of the emblems.

James J. Modarelli, head of the Research Reports Division at the NASA Lewis Research Center (now the NASA Glenn Research Center), was the chief designer of the NASA seal and meatball insignia.[1] The story of Modarelli's conceptualization and involvement in the creation of the logos is a key part of this publication.

Modarelli's designs evolved and changed as the process continued. He was assisted by the Heraldic

---

1. Joseph Chambers, "The NASA Seal and Insignia, Part 1," *NASA History News & Notes,* Volume 30, Number 2, 2013. pp. 11–15.

Branch of the Army Office of the Quartermaster General (now the Army Institute of Heraldry) in the development of the final emblems and the presidential approval process.

THE VISION BEGINS

In July 1958, Modarelli attended the triennial inspection of the AAL, during which facilities and research efforts within the NACA were highlighted and discussed for invited guests in the scientific community. At the time, the annual NACA inspections rotated between the Ames, Langley, and Lewis laboratories

every 3 years. The host laboratory provided the majority of the briefings, while the other two labs provided a few exhibits and presentations on additional topics.[2] The emphasis of the 1958 program was on space-related research, intended to display to visitors the emerging NACA activities in the critical disciplines.

During the Ames meeting, Modarelli participated in a tour consisting of nine stops for presentations on topical research activities. At the Ames Unitary Plan Wind Tunnel, he viewed an Ames exhibit featuring a discussion by Ames researchers on current advanced supersonic aircraft technology. On display was an Ames wind-tunnel model of a radical supersonic airplane configuration designed for efficient flight at Mach 3. Featuring a cambered and twisted arrow wing with an upturned nose, the sleek model deeply impressed Modarelli as a symbol of the leading-edge aeronautical efforts of the NACA.[3]

After the Ames inspection, Modarelli returned to Lewis where he continued to participate in technical conferences and visits to other NACA laboratories. During a visit to Langley in late summer, he saw another highly swept and cambered arrow wing model that had been tested in the Langley Unitary Tunnel at supersonic speeds.[4] The efforts at both Ames and

FIGURE 3-1.

James J. Modarelli (right) of the NASA Lewis Research Center (now the NASA Glenn Research Center) was the designer of the NASA seal and insignia. In the picture, Modarelli presents an incentive award to an employee at Lewis in 1964. (NASA C-68827)

2   National Archives and Records Administration, San Bruno, CA. Record Group 255.4.1. The nine-stop program included seven exhibits from Ames (Earth Satellites, Aerophysics, Aerodynamic Heating, Hypervelocity and Entry Research Techniques, Stability During Atmosphere Entry, Piloting Problems During Entry, and Supersonic Airplanes). A Lewis exhibit was on Space Propulsion Systems, and a Langley exhibit was on Flight Research for Space Craft.

3   James J. Modarelli, interview by Mark A. Chambers, 17 July 1992. Chambers, Mark A., "History of the Red 'V' in the NASA Meatball," NASA Headquarters Historical Records Collection, File 4542, 31 July 1992.

4   James J. Modarelli, interview by Mark A. Chambers, 17 July 1992.

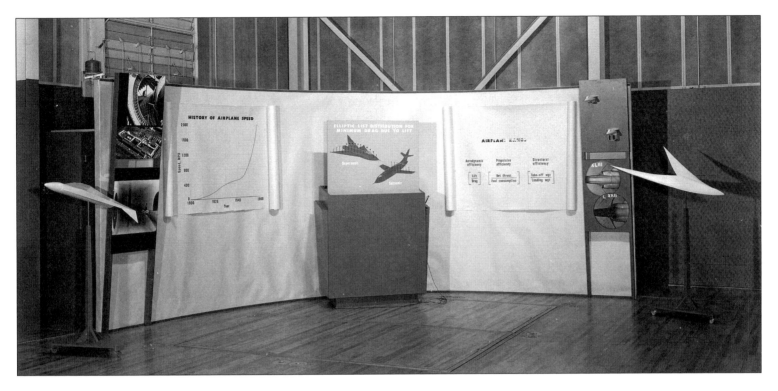

FIGURE 3-2.

View of the 1958 triennial inspection display on supersonic airplanes that was viewed by James Modarelli in the Ames Unitary Plan Wind Tunnel High Bay. The large Ames model of a twisted and cambered supersonic wing configuration at the right side of the picture impressed Modarelli and inspired him to use the concept in the NASA logos. (NACA A-24000-5H (2))

Langley had been stimulated by the pioneering efforts of the brilliant NACA/NASA scientist R. T. Jones in his quest for aerodynamic efficiency at supersonic speeds. Jones is recognized as the first American scientist to analyze and advocate for the beneficial effects of wing sweep on high-speed flight. When Jones transferred from Langley to Ames in 1946, he retained his interest in the subject, maintaining communication with peers at Langley while initiating new research at Ames.[5]

The leader of research under Jones at Ames was Elliott D. Katzen, while Clinton E. Brown led the work at Langley. In 1958, after reviewing the advanced wings created by both groups, Modarelli stylized the radical features of the arrow-wing configuration in his evolution of the NASA seal design. The wing would also ultimately become an element of the NASA insignia.[6]

5   Interview of R. T. Jones by Mark A. Chambers, 24 July 1992. Chambers, Mark A., "History of the Red 'V' in the NASA Meatball," NASA History Office File 4542, 31 July 1992.

6   James J. Modarelli, interview by Mark A. Chambers, 17 July 1992; J. C. South, Jr., "Meatball Logo Based on Wind Tunnel Model," *The Researcher News*, in-house newsletter, NASA Langley Research Center, 14 August 1992, p. 2.

The NASA Seal

The creation of an official NASA seal and insignia was initiated in mid-September 1958 when Victory, then Executive Secretary of the NACA, sent letters to the Ames, Langley, and Lewis laboratories soliciting suggestions for an insignia design for the new organization.[7] In addition to the request to the laboratories, an ad hoc committee of staff members at NACA Headquarters was invited to submit candidate designs.[8] A few weeks after the invitation for designs was sent, NASA absorbed the personnel, facilities, and research activities of the NACA at the close of business on 30 September.

Modarelli was well prepared for the design competition. He had already selected the twisted and cambered arrow-wing concept that he had seen at Ames and Langley earlier in the year as an element he would use in his seal design to represent NASA's aeronautics program. Modarelli also discussed a competing seal design with Harry J. DeVoto, head of the Ames Graphics and Exhibits Branch. DeVoto's design included the traditional circular shape required for all government seals, featuring an outer circle enclosing the words "NATIONAL AERONAUTICS AND SPACE ADMINISTRATION" and an inner circle.

Inside the inner circle was a blue field with several star shapes, a globe representing Earth in the center of the blue field, and the path of a nascent orbiting body circling the globe. Modarelli adapted the DeVoto design for the Lewis submittal by adding the advanced supersonic wing and by modifying the globe and orbiting body.[9] The stars in the design were not representative of specific constellations, but were artistic creations. Likewise, the orbiting body was not based on a specific satellite.

Members of the illustration section of the Research Reports Division at Lewis worked on candidate designs for the seal under the direction of Modarelli.[10] His personal design, however, received most of the attention.

The ad hoc committee at Headquarters completed designs of their candidates for the NASA insignia and submitted them to Victory on 6 October 1958, in advance of the Center submissions.[11] The four designs were simplistic variations of the letters "N-A-S-A" encircled by an orbiting body. The committee's recommended design included these elements and a shock wave superimposed to represent aeronautics. NASA's first Administrator, Dr. T. Keith Glennan, rejected those designs during the selection process.

After the seal design candidates from the Centers had been submitted to Administrator Glennan for final selection, several members of Glennan's staff met

---

7   "Staff Invited to Submit Designs for NASA Insignia," Langley *Air Scoop*, in-house newsletter, NACA Langley, 19 September 1958. Volume 17 Issue 38, p. 2. Also, *The Orbit*, in-house newsletter, NACA Lewis Research Center, 30 September 1958, 1. In reality, the invitation was to design candidates for the NASA seal. No such invitation occurred for the NASA insignia.

8   Memorandum from Patrick A. Gavin to John F. Victory on Proposed NASA Insignia, 6 October 1958. NASA Headquarters Historical Records Collection, file 4540. Members of the Headquarters committee were Patrick Gavin, Paul Dembling, Robert Lacklen, and E. O. Pearson.

9   "Ames Research Center View of NASA Logo Design," Letter from Harry J. DeVoto to Steve Garber, 6 May 2001. Headquarters Historical Records Collection, file 4540.

10   "This Is NASA Insignia," *The Orbit,* in-house newsletter, NASA Lewis Research Center (31 July 1959), p. 1. Lewis illustrators Richard Schulke, Louise Fergus, and John Hopkins assisted Modarelli in the design effort.

11   Gavin Memorandum to John F. Victory. 6 October 1958. NASA Headquarters Historical Records Collection, file 4540. The file includes very poor copies of the designs.

with him in December 1958 to decide the winner. After considerable discussion about the candidate designs, Glennan declared that they were "wasting time," and he personally chose the seal design submitted by Modarelli.[12] As will be discussed, a major artistic error in the layout of the selected emblem had been made unknowingly during the design process and would ultimately require a revision. In particular, the red wing element had been rendered in an upside-down attitude. Despite the graphic error, however, the initial design moved forward through the process for presidential approval.

The reason for the upside-down rendering of the wing is not known. However, several photographs of cambered and twisted arrow-wing models were taken at the Centers and submitted to Modarelli at his request for guidance in the design process. At least one of the photos showed a model in an inverted pose that could have been misinterpreted by an artist. A wooden model of Clint Brown's wing had been constructed to show the unique features of the configuration. On 28 January 1958, photographs had been made which included the model in an inverted pose on a table. The picture might have caused the misinterpretation shown in the graphic.

FIGURE 3-3.

Sketches of the six competing Center designs for the NASA seal. The winning design submitted by James Modarelli and his Lewis team is at the lower right. Note the upside-down attitude of the wing element. (NASA Headquarters Historical Reference Collection (HRC), file number 4540)

---

12 Letter from Dr. T. Keith Glennan to Daniel Goldin, Administrator of NASA, 23 June 1992 (NASA Headquarters HRC File A92-1899). In the letter Dr. Glennan congratulated Goldin on bringing back the insignia designed by James Modarelli and recalled his personal actions at the selection meeting. Dr. Glennan served as NASA Administrator for 30 months. For his biography and video of an extended interview, see *http://www.ieeeghn.org/wiki/index.php/Keith_Glennan* (accessed 18 September 2014). Also see J. D. Hunley, editor, *The Birth of NASA: The Diary of T. Keith Glennan*. NASA SP-4105, 1993.

FIGURE 3-4A.

A three-quarter rear view of a wooden Langley display model in January 1958 showing the radical twist and camber of a supersonic arrow-wing design. Note the cobra-like raised nose at the upper right and the cambered, drooped trailing edges of the 75-degree swept wing. These features were inverted in the first seal design by Modarelli. (NACA L-00502)

FIGURE 3-4B.

This photograph of the inverted model could have resulted in the erroneous wing rendering in Modarelli's first design of the NASA seal later in 1958. (NACA L-00504)

THE PATH TO PRESIDENTIAL APPROVAL

Acquiring presidential approval for the NASA seal involved a mandatory four-step process for all government agencies. First, the Heraldic Branch of the Army Office of the Quartermaster General (now the Army Institute of Heraldry) had to be contacted for assistance with the design of the seal as well as with other stages of the presidential approval process.[13] In addition to analysis and recommendations for the proposed graphics, the Heraldic Branch would prepare plaster reproductions of the proposed seal and complete other tasks as required. After the seal design was finalized to the satisfaction of NASA and the Heraldic Branch, the Commission of Fine Arts would review its artistic merits. In 1921, President Warren Harding had delegated this type of responsibility to the fine arts commission.[14]

Following approval by the commission, the process required the NASA Administrator to approve the design and submit a formal request to the Bureau of Budget for presidential approval. When all of this was complete, a Presidential Executive Order would be signed, officially establishing the seal. On 23 January 1959, a month after Administrator Glennan had

13 The Army Institute of Heraldry has a long and distinguished record of support to the United States Army. In 1919, President Woodrow Wilson directed the creation of the Heraldic Program Office under the War Department General Staff. Its purpose was to take responsibility for the coordination and approval of coats of arms and other insignia for Army organizations. By the end of WWII, its role expanded to include the other military services. In 1957, Public Law 85-263 directed the Secretary of the Army to furnish heraldic services to all branches of the federal government. The Institute's wide range of heraldic services includes decorations, flags, streamers, agency seals, coats of arms, badges, and other forms of official emblems and insignia.

14 Kohler, Sue A. *The Commission of Fine Arts: A Brief History, 1910–1995.* Washington, DC: United States Commission of Fine Arts, 1996, p. 204.

FIGURE 3-5.

President Dwight D. Eisenhower (center) presents commissions to Dr. T. Keith Glennan (right) as the first Administrator of NASA and Dr. Hugh L. Dryden (left) as Deputy Administrator. (NASA MSFC-9248169)

selected the Modarelli seal design, NASA asked the Heraldic Branch to initiate the review process, which included a presentation to the Commission of Fine Arts.[15] The Modarelli design survived the Army review with only minor changes in color shades.

The Quartermaster Corps sent the design to the Commission of Fine Arts on 17 February for comments and approval.[16] The fine arts commission responded the following week by "reluctantly" approving the design, stating, "This design is primarily a pictorial conception and is very reminiscent of designs appearing in current commercial advertising." The commission also asked that the letters "U.S.A." at the bottom of the name band be enlarged to the size of the other letters and that the stars in the name band be deleted.[17] NASA accepted these changes on 24 March, instructing the Army to proceed with the development of drawings and a plaster model of the design seal, with the modifications suggested by the arts commission. The statement also requested that the Army take

15 Robert J. Lacklen, NASA Director of Personnel, letter to Lt. Gen. J. F. Collins, Deputy Chief of Staff for Personnel, Office of Secretary for the Army, 23 January 1959. Files of the Institute of Heraldry, Department of the Army, Fort Belvoir, VA, to be deposited in the NASA HRC.

16 Letter to the Commission of Fine Arts, Interior Department Building, Washington, DC, from Lt. Col. James S. Cook, Jr., Chief, Heraldic Branch, Quartermaster Corps, 17 February 1959. Files of the Institute of Heraldry, Department of the Army, Fort Belvoir, VA, to be deposited in the NASA HRC.

17 L. R. Wilson, Secretary of the Commission of Fine Arts, letter to Lt. Col. James S. Cook, Jr., Chief, Heraldic Branch of the Army Office of the Quartermaster Corps, 25 February 1959. Files of the Institute of Heraldry, Department of the Army, Fort Belvoir, VA, to be deposited in the NASA HRC.

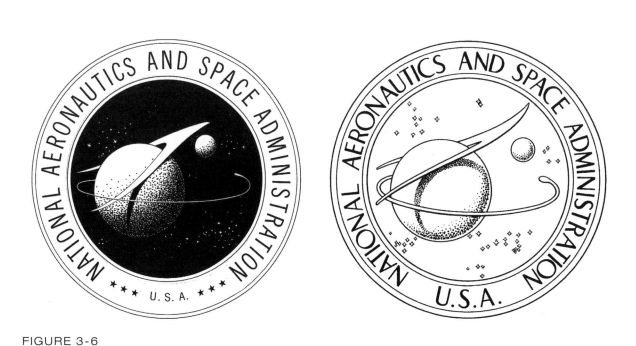

FIGURE 3-6

Sketch on the left shows the erroneous NASA seal originally sent to the Army Heraldic Branch and the Commission of Fine Arts for review. Sketch on the right shows the corrected version in July. Note the revised wing attitude and shape, omission of the stars, enlargement of the letters "U.S.A." in the border, altered star patterns, and modified shadow pattern on the globe. (Institute of Heraldry, Department of the Army)

appropriate steps to have the seal officially established by executive order of the President.[18]

After the design go-ahead had been submitted to the Quartermaster Corps—and approved by the Commission of Fine Arts—copies of the proposed seal were sent to the NASA Centers for the first time for information in late March. When the design layout reached Langley, Clint Brown took one look and remarked, "They've drawn my wing upside down!" and he quickly pointed out the mistake to management.[19]

When word of the mistake reached NASA Headquarters in late March, the work by the Heraldic Branch was suspended while Modarelli corrected the wing problem and made other changes to the graphic.[20] In addition to depicting the wing in the proper

---

18 Robert J. Lacklen, Director of Personnel, NASA, letter to Lt. Col. James S. Cook, Jr., Chief, Heraldic Branch of the Army Office of the Quartermaster Corps, 24 March 1959. Files of the Institute of Heraldry, Department of the Army, Fort Belvoir, VA, to be deposited in the NASA HRC.

19 Langley retiree Jack Crenshaw, interview by Joseph Chambers, 1 February 2013.

20 Memorandum for record, telephone call to Ms. Landrum, Heraldic Branch of the Army Office of the Quartermaster Corps, from Mr. Gavin, NASA Headquarters, 31 March 1959. Gavin called to suspend all action on the seal pending further advice from NASA. Files of the Institute of Heraldry, Department of the Army, Fort Belvoir, VA, to be deposited in the NASA HRC.

erect attitude, the shape of the wing was changed to that of "a theoretical wing configuration." The sweep angle of the wing was increased, the entire width of the right wing was reduced to give an impression of an edge-on view, and the tail of the left wing was contoured to suggest a continuous sweep to the tip. The shadow the wing cast on the globe element was modified to conform to the shadow effect obtained from photographs of models of Earth and the wing. Photographs of the shadow effect were forwarded to the Heraldic Branch for guidance. The height of the small blue sphere was reduced and the color of the shadow was changed to match the pattern used on the yellow sphere. Alterations were also made to the elliptical path of the space vehicle.[21] Finally, the stars within the field were modified in size and location, and the font used for the letters in the border was changed. The details of the graphic continued to change as the final seal design evolved.

On 21 May 1959, a NASA-Army conference was held to discuss all aspects of the requested Army actions. The topics for discussion included the seal (already completed); a plaster model, mold, and line drawing of the seal; design of a decal-type version of the seal; design of an Agency flag for NASA; design for a lapel pin based on length of service; and design of the NASA Exceptional Service and Distinguished Service Medals, including suspension ribbon, metal pendants, and lapel emblems.[22]

FIGURE 3-7.

Final version of the
NASA seal.

The Army Heraldic Branch continued to work on the NASA request through June and into early July. Modarelli and others from NASA Headquarters were active participants with the Army during the process. The revisions to the seal design were completed by Army and NASA participants in late July. The modified design sent to the Administrator for approval differed greatly from the original design that had been submitted to and approved by the Commission of Fine Arts in February. Administrator Glennan approved the final design for the NASA seal in August.[23]

NASA Headquarters informed the field Centers that more than 350 designs had been submitted for consideration by staff members from field Centers and Headquarters.[24] According to the announcement, the

---

21 Robert J. Lacklen, Director of Personnel, NASA, letter to Lt. Col. James S. Cook, Jr., Chief, Heraldic Branch of the Army Office of the Quartermaster Corps, 13 April 1959. Files of the Institute of Heraldry, Department of the Army, Fort Belvoir, VA, to be deposited in the NASA HRC.

22 Memorandum for record, Ms. Landrum, Heraldic Branch of the Army Office of the Quartermaster Corps, 21 May 1959. Attendees at the meeting included Mr. Gavin and Mr. Maloney of NASA and Mr. Potter and Ms. Landrum of the Heraldic Branch. Files of the Institute of Heraldry, Department of the Army, Fort Belvoir, VA, to be deposited in the NASA HRC.

23 Patrick A. Gavin, Acting Director of Personnel, NASA, letter to Lt. Col. James S. Cook, Jr., Chief, Heraldic Branch of the Army Office of the Quartermaster General, 14 August 1959. Files of the Institute of Heraldry, Department of the Army, Fort Belvoir, VA, to be deposited in the NASA HRC.

24 "Insignia Incorporates Features contained in Number of Suggestions," Langley *Air Scoop*, in-house newsletter, 31 July 1959, Volume 18, Issue 30. p. 1.

final selection was not a copy of any one submittal, but was developed by incorporating features contained in a number of candidate designs. As a result, no individual award or special recognition was made under the NASA Incentive Award Plan.

President Dwight D. Eisenhower signed Executive Order 10849 establishing the seal on 27 November 1959. The order included a description of the seal:

> On a disc of the blue sky strewn with white stars, to dexter a large yellow sphere bearing a red flight symbol apex in upper sinister and wings enveloping and casting a gray-blue shadow upon the sphere, all partially encircled with a horizontal white orbit, in sinister a small light-blue sphere; circumscribing the disk a white band edged gold inscribed "National Aeronautics and Space Administration U.S.A." in red letters.

The seal was later amended early in the Kennedy administration when the color of the shadow on the sphere was changed from gray-blue to brown under Executive Order 10942, 22 May 1961.

The NASA Insignia

Dr. Glennan, NASA's first administrator, asked James Modarelli to design a simplified insignia for the informal uses of the new Agency. The insignia would appear on items such as lapel pins and signs on buildings and facilities. Modarelli worked on the insignia design task while completing the seal design. He chose the main elements from the seal for the simplified insignia— the circle, representing the planets; stars, representing space; the advanced supersonic wing, representing aeronautics; and an orbiting spacecraft. He then added

FIGURE 3-8.

The NASA insignia designed by James Modarelli.

the letters "N-A-S-A."[25] The colors of the insignia are Pantone 185 (red) and Pantone 286 (blue).[26]

In April 1959, NASA formally notified the Heraldic Branch of the Army Office of the Quartermaster General that it would use the insignia design created by Modarelli and that NASA would not require its services for this undertaking.

Dr. Glennan announced the new NASA insignia in the NASA Management Manual on 15 July 1959.[27]

---

25 "This Is NASA Insignia," *The Orbit*, in-house newsletter at Lewis Research Center (31 July 1959), p. 1. Lewis illustrators Richard Schulke, Louise Fergus, and John Hopkins assisted Modarelli in the design effort.

26 NASA Style Guide, November 2006. p. 8.

27 The insignia had been approved by then Administrator Glennan before he approved the seal. The announcement appeared in several in-house newsletters: "Cite Regulations for use of New NASA Insignia," *Air Scoop*, newsletter at Langley Research Center (24 July 1959); "This Is NASA

Specifications and orders for insignia lapel pins, decals, and stationery were completed during the fall of 1959. In November 1959, Modarelli transferred to NASA Headquarters, where he held the position of Director of Exhibits for NASA until he returned to Lewis Research Center in 1961.[28]

Modarelli's creation would serve as the official NASA insignia for 16 years before being replaced in 1975; it would later be reinstituted in 1992.

THE RED WING

At this juncture, it might be informative to the reader to clarify the oft-misunderstood red symbol that Modarelli chose to represent the aeronautics thrust of the Agency in the NASA seal and insignia. As previously discussed, the symbol was a stylization of an actual NACA-designed wing concept for supersonic applications involving flight at Mach 3. This section explains why the supersonic-wing model was part of the display at Ames and also discusses the individuals and technical programs at Ames and Langley that led to the concept.

In the mid-1950s, the growing interest in high-speed flight resulted in accelerated efforts to maximize the efficiency of wings at supersonic speeds. A wide variety of wing shapes were investigated by the NACA in both experimental and theoretical studies at the NACA Langley Aeronautical Laboratory and at the NACA AAL.[29] Conclusions from these studies indicated that arrow wings promised superior performance at supersonic speeds compared to delta wings and other shapes. Furthermore, the research showed that the lift-drag characteristics of arrow-wing configurations could be appreciably enhanced by cambering and twisting the wings. Aerodynamicists under the direction of John P. Stack at Langley and Robert T. Jones at Ames were major contributors to the research and were recognized as world leaders in theoretical and experimental methods for supersonic aerodynamics.

The leaders of research on the design of twisted and cambered supersonic arrow wings were Elliott D. Katzen at Ames and Clinton E. Brown and Francis E. McLean at Langley. The emphasis in Katzen's research was on how to optimize the performance of the wings at supersonic speeds.[30] At Langley, Brown and McLean were primarily interested in the practical applications of the wing concept to realistic airplane configurations with satisfactory stability and performance for specific missions.[31] These simultaneous projects were typical of productive NACA/NASA efforts in certain research areas. That is, once the committee identified the importance of a topic, the various labs would undertake different approaches to assess the potential benefits of multiple concepts, identify and resolve problems, and create valid design methods for applications of the concept. Efforts were carefully planned to avoid unnecessary duplication of efforts between Centers.

Insignia," *The Orbit*, newsletter at Lewis Research Center (31 July 1959); "New NASA Insignia Approved," *The Astrogram*, newsletter at Ames Research Center (6 August 1959).

28 "Modarelli Transfers to Headquarters," *The Orbit*, in-house newsletter at Lewis Research Center (20 November 1959), p. 1.

29 F. Edward McLean: *Supersonic Cruise Technology*, NASA SP-472, 1985.

30 An example of Katzen's research: Elliott D. Katzen, "Idealized Wings and Wing-Bodies at a Mach Number of 3," NACA Technical Note 4361, July 1958.

31 An example of Brown's research: Brown, Clinton E.; McLean, F. E.; and Klunker, E. B.: "Theoretical and Experimental Studies of Cambered and Twisted Wings Optimized for Flight at Supersonic Speeds." Presented at the Second International Congress of the Institute of the Aeronautical Sciences, Zürich, Switzerland, 12–16 September 1960.

FIGURE 3-9.

Elliott Katzen of the NASA Ames Research Center. (Photo courtesy of Mrs. Elliott Katzen)

FIGURE 3-10.

Clint Brown of the NASA Langley Research Center. (NASA L-05232)

By the time of Modarelli's visit to Ames in 1958, several notable events had occurred. Beginning in the early 1950s, both Ames and Langley had explored theoretical analyses of the aerodynamic benefits of the advanced wing concepts. Jones and Brown had exchanged ideas and research plans on the design of the advanced wing concept. Brown designed a wing layout for an initial theoretical analysis, the results of which he shared with Jones. Jones subsequently used the results in planning the Ames program.[32] The Ames work focused on wings with a leading-edge sweep of

32   Second Interview of R. T. Jones by Mark A. Chambers, 27 July 1992. Chambers, Mark A.: "History of the Red 'V' in the NASA Meatball," NASA History Office File 4542, 31 July 1992. Also, letter to James Modarelli from Mark Chambers requesting comments on arrow wings tested at Ames and Langley, 31 July 1992.

80 degrees, whereas the Langley efforts were dominated by wings with a sweep angle of 75 degrees.

Katzen conducted tests on a series of twisted and cambered arrow wings with various wing sweep angles and Mach numbers in the Ames 1- by 3-Foot Supersonic Tunnel. With guidance from Jones, he began studies to optimize arrow-wing shapes in 1955. After conducting theoretical analyses, he followed up with wind-tunnel tests in 1955 and 1956 to determine which of his designs had the highest value of lift-drag ratio at Mach 3.

Jones had suggested to Katzen that a wing sweep of about 80 degrees would be a nearly optimum shape for a Mach 3 application. For that condition, the leading edge of the wing would be behind the Mach cone, and the Mach number perpendicular to the leading edge in flight at Mach 3 would be comparable to that of a Boeing 707 subsonic transport in flight. He also suggested using a Clark-Y airfoil with camber but no twist. When the first thin metal wing arrived from the shop, Katzen took it to Jones for inspection and Jones twisted it to the shape he thought would be the best performer. Subsequent tunnel tests in the Ames 1- by 3-Foot Supersonic Tunnel indicated a maximum lift-drag ratio of 9—the highest efficiency ever measured at the time for an isolated wing designed for Mach 3.[33] Analysis of approaches to enhanced performance at Mach 3 suggested that a lifting forebody should be included in the configuration, resulting in the peculiar "upturned nose" seen in the NASA insignia and seal.[34]

Observers that were in the office with Katzen when he first saw the design of the NASA seal reported that he was terribly excited—pointing out that, in his opinion (and those of the observers), the red wing on the

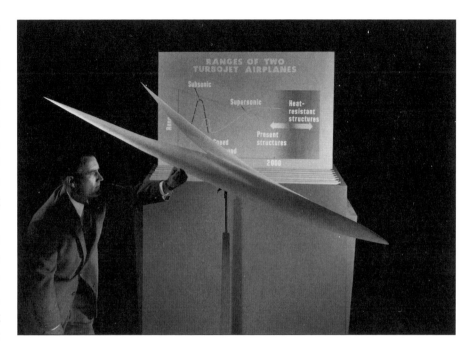

FIGURE 3-11.

Ames researcher Lynn W. Hutton poses with an advanced supersonic wing model in August 1958. This particular model was the one on display at the Ames inspection the previous month. (NACA A-24208)

seal depicted his wing design. Katzen, however, was never given credit by the Agency as the designer of the red wing in the NASA insignia, and he became deeply upset over the lack of recognition.[35] When Katzen passed away on 12 February 2012, the memorial pamphlet at his funeral mentioned that he and Jones had perfected a "highly swept wing design that was actually

33 Bugos, Glenn E.: *Atmosphere of Freedom: Sixty Years at the NASA Ames Research Center* (NASA SP-4314, 2000), p. 36.

34 Brown, McLean, and Klunker, p. 13.

35 William C. Pitts, interview by Joseph Chambers, 5 December 2012. Pitts was a member of the technical organization of Elliott Katzen under Wallace F. Davis at Ames. Pitts was Katzen's officemate and witnessed the events cited. Noted aerodynamicist Jack Neilsen was also a member of the group aware of Katzen's contribution.

represented as the red triangle dart across the NASA meatball logo."[36]

At Langley, development of the wing concept was spurred on by an evolving military need for flight at Mach 3. As tensions between the United States and the Soviet Union increased during the 1950s, one of the highest priority projects for the U. S. Air Force was developing an intercontinental supersonic bomber with an unrefueled combat radius of 4,000 nautical miles and the capability to deliver nuclear weapons to the Soviet Union. In 1955, the Air Force initiated Weapon System 110A (WS-110A) for industry design studies that ultimately led to the North American XB-70 and its demonstrated Mach 3 capability.

In view of the criticality of supersonic efficiency to the WS-110 mission, a special classified study was led by Langley to assess the aerodynamic benefits of a 75-degree twisted and cambered arrow-wing bomber designed by Brown and McLean. The wing was contoured to provide trim and stability at the design condition for the WS-110A mission, including the necessary internal volume for fuel and weapons. The configuration was referred to within Langley as the "Brown Bomber."

The testing effort included evaluations of the Brown-McLean design in the Langley Unitary Plan Wind Tunnel in 1957 and 1958 and the Langley 7- by 10-Foot High-Speed Tunnel in 1958.[37] The objective of the Unitary tests was to measure aerodynamic performance at Mach 3, and the 7- by 10-Foot Tunnel test was to determine the low-speed landing, takeoff, and subsonic performance and stability characteristics of the configuration.[38] Both tests included the isolated wing and addition of components necessary for a real airplane including tail surfaces, engine pods, etc.

Although the results of the supersonic wind-tunnel tests at Langley indicated an impressive maximum lift-drag ratio of 6.0 for the isolated wing at Mach 3, Brown was extremely disappointed because this value was appreciably lower than that anticipated based on his theoretical studies. Flow visualization studies revealed that flow separation on the upper wing surface occurred that could not be predicted using the limited theories of the day.

The results of the research programs on the advanced wing concepts at Ames and Langley were reported in a session at the NACA Conference on High-Speed Aerodynamics held at Ames 18–20 March 1958. The primary purpose of the conference was to convey to the military services and their contractors the results of recent supersonic research and to provide those in attendance with an opportunity to discuss the results. Katzen reported on his efforts at Ames, and Donald D. Baals presented the results of the Langley activities.[39, 40]

36 Funeral pamphlet for Elliott D. Katzen, to be deposited in the NASA HRC. Also see the announcement of Katzen's passing in the Ames *Astrogram*, Spring 2012, p. 6.

37 Entered in test log of Langley Unitary Tunnel Test Section I as "Test 70 WS-110A Brown Bomber in UWT Low-Mach Number Test Section," Langley Historical Archives, *Wind-Tunnels* by Donald Baals. Results of two supersonic tunnel entries are reported in Hallissy, Joseph M., Jr., and Hasson, Dennis F.: *Aerodynamic Characteristics at Mach Numbers 2.36 and 2.87 of an Airplane Configuration having a Cambered Arrow Wing with a 75° Swept Leading Edge*, NACA RM L58E21, 4 August 1958.

38 The subsonic tests are discussed in Davenport, E. E. and Naseth, R. L., *Low-Speed Wind Tunnel Investigation of the Aerodynamic Characteristics of an Airplane Having a Cambered Arrow Wing with a 75° Swept Leading Edge*, NASA TM X-185, 1960.

39 Baals had a long and active career at the NACA and NASA, having designed Langley's 4 by 4-Foot Supersonic Pressure Tunnel in the 1940s. He continued his career at Langley after the Center became a part of NASA and retired in 1975.

40 Elliott D. Katzen, "Idealized Wings and Wing-Bodies at a Mach Number of 3," Paper 38, pp. 509–520; and Donald D. Baals, Thomas A. Toll, and Owen G. Morris, "Airplane Configurations for cruise at a Mach Number of 3," Paper 39, pp. 521–542.

FIGURE 3-12.

The Brown Bomber configuration during tests in the Langley Unitary Plan Tunnel in February 1958. (NASA LAL 58-826)

After testing of the NACA Brown Bomber was completed at Langley in 1958, a red plastic display model was made from a mold of the wind-tunnel model. The red model was on display for visitors in Brown's office and was then exhibited at the Unitary Plan Wind Tunnel for many years.[41]

In summary, the evidence gathered in this study—based on interviews of James Modarelli—reveals that he was inspired to include the arrow wing shape in the NASA logos by his encounter with the model during the Ames inspection. He then became aware of similar models at Langley and used common features to create his own artistic interpretation of a representative wing used in the NASA emblems. As previously discussed, his wing design changed as the interactions with the Army Heraldic Branch proceeded. A fitting conclusion is that all three Centers and NASA Headquarters played a role in the creation of the NASA seal and insignia.[42]

The U.S. Air Force and its industry teams that were part of the WS-110A project rightfully considered the highly swept cambered arrow-wing concept an unacceptably immature approach that would require years of extensive research in a number of disciplines, including propulsion integration, aeroelasticity, and flight control. Nonetheless, the lessons learned from wind-tunnel testing and theoretical analyses by Katzen, Brown, and McLean provided building blocks that later played key roles in NASA research on civil high-speed transports. The superior performance of the arrow wing for supersonic missions continues to make it a strong contender for future supersonic transport designs.[43]

41 Jerry C. South, Jr., "Meatball Logo Based on Wind Tunnel Model," *The Researcher News,* in-house newsletter, NASA Langley Research Center, 14 August 1992, p. 2. Also, Roy V. Harris, Jr., interview by Joseph Chambers, 1 November 2012.

42 Letter from Harry J. DeVoto, retired Chief of Graphics and Exhibits at NACA Ames, to Daniel S. Goldin, NASA Administrator, 11 June 1992. Headquarters Historical Records Collection, file 4540.

43 F. Edward McLean, Supersonic Cruise Technology, NASA SP-472, 1985.

Applications During Transition

As might be expected in a time of rapid and dramatic change, questions regarding the new markings required on property owned by NASA quickly surfaced. In many cases, the Centers proceeded with their interpretation of appropriate markings and emblems. At the very minimum, an effort had to be made to change the letter C in NACA to an S. Perhaps the greatest effort was repainting the roofs of the Agency's flight hangars, since they had been painted with prominent NACA markings as previously discussed. In addition, the entire NACA research aircraft fleet had been emblazoned with the NACA wings emblem on their tail bands.[44]

---

44 On most aircraft, the NACA letters were changed to NASA, while a few had the tail band removed.

FIGURE 3-13.

In this very rare picture, an F-86D flown by Langley pilots during the transition from the NACA to NASA in 1958 was marked with an unofficial modification to the traditional NACA wings insignia. A NASA designation appears in the wings insignia as well as on the aft fuselage.

FIGURE 3-14.

Tradition changes as the letter C in the NACA logo on the roof of the hangar at the Lewis Research Center is removed. Note the existing NACA emblems above the hangar doors and on the fuel tanker. (NASA C-1958-48854)

FIGURE 3-15.

In this photo, the Lewis hangar roof and entrance markings have been updated and the F-94 research aircraft has been marked with NASA on the forward fuselage. The old NACA wings emblem, however, has been marked over with "NASA."

FIGURE 3-16.

Workmen begin the task of replacing NACA emblems and signs by removing the NACA wings insignia over the entryway at the High-Speed Flight Station. (NASA E96-43403-4)

CHAPTER 4

Fly Me to the Moon, 1960–1974

Insignia and Seal: Onward and Upward

Led by an enthusiastic and dedicated new Agency, the United States took on the scientific challenges and potential security threats posed by the Soviet Union with an aggressive space program. President John F. Kennedy's directive to send Americans to the Moon within the decade brought forth unprecedented and exciting accomplishments that ultimately met his goal. The new NASA seal and insignia became iconic fixtures of the space program and were displayed with pride throughout the Agency. The insignia, in particular, was displayed at virtually every major event of the period—to the point where it often was referred to as the NASA seal. The confusion was caused by the almost unnoticed use of the seal on the NASA flag or on the rostrum for presentations by the Administrator.

During an era in which NASA seemed to accomplish the impossible and generate unbridled support from the U.S. public, the logos created by Modarelli remained unaltered from his designs of 1959. Those logos would continue to grow in popularity and advance international recognition of the world-class organization that had led the nation to the Moon while generating innumerable civil and military applications of the technologies it conceived, developed, and matured.

CRITICISM

Despite widespread acceptance and support for the NASA logos, a few incidents arose in which representatives of the art world were critical of Modarelli's designs. A specific example involved one of NASA's highest honors, the NASA Distinguished Service Medal, during the early days of Project Mercury.

At the request of NASA, several proposed designs for the Distinguished Service Medal and Exceptional Service Medal had been prepared by the Heraldic Services Division of the Army and submitted to NASA in early December 1959. The Army had also submitted designs for the suspension ribbons, rosettes, and lapel emblems for the medals. In January 1960, NASA advised the Army of the design that had been selected and also informed them that the development and

FIGURE 4-1.

NASA's top management from 1958 to 1960 was Dr. T. Keith Glennan, Administrator (center); Hugh L. Dryden, Deputy Administrator (left); and Richard E. Horner, Associate Administrator (right). Taken on 1 March 1960, the picture shows the NASA seal above Administrator Glennan and the NASA flag behind Dryden. (NASA 60-ADM-7)

production of medals and ribbons would be handled by the Agency with no further Army responsibilities. The NASA medals designed by the Army had the same obverse (front) as the NASA seal.[1]

After the inspiring suborbital flight of astronaut Alan Shepard on 5 May 1961, President Kennedy awarded him the NASA Distinguished Service Medal in recognition of his brave and exceptional feat. In the wake of this event, however, the first critical comments regarding the design of NASA logos surfaced.

In its 19 May 1961 issue, *Time* magazine published a criticism of the design of government medals in general, with emphasis on the design of the NASA Distinguished Service Medal:

> Since the days of ancient Athens, a brave act has deserved a proud and artistic medal—everywhere but in the U.S. Last week when President Kennedy honored the country's first astronaut, all he had to pin on the lapel of Commander Alan Shepard was something that looked as if it might have come out of a Cracker Jack box. The Distinguished Service Medal of the National Aeronautics and Space Administration is the most unimaginative decoration turned out by the U.S. government so far—and the competition for that title is stiff…. Such U.S. medals are turned out by the U.S. Army Institute of Heraldry, which has only two trained sculptors on its staff…. No one at NASA will say

1   Fact sheets prepared by Col. John D. Martz, Jr., Commander of Quartermaster Corps, on background of Heraldic Services Division participation in design of NASA seal and medals, 18 May 1961. The fact sheets were prepared at the request of higher-level officials within the Army. Files of the Institute of Heraldry, Department of the Army, Fort Belvoir, VA, to be deposited in HRC.

who is responsible for Commander Shepard's D.S.M., but that perhaps is a blessing. One side of the medal shows a planet and satellite—a motif that any schoolboys might have thought up. On the other side is the inevitable laurel wreath. As for the lettering, Designer Henry Hart of the Smithsonian Institution has just one word: "Atrocious."[2]

As expected, many in the NASA community reacted adversely to the article. Harry J. DeVoto, who had been Head of the Graphics and Exhibits Branch at NASA Ames, was particularly offended. DeVoto had been the designer who influenced Modarelli's choice of the globe and orbiting spacecraft in the logos. In 2001, DeVoto commented:

> History has proven *Time* magazine's assessment of the logo entirely wrong. In an issue that was published about the time the logo was made public, *Time* used words like, "Cracker-Jack toy", "typical government art" and many other negative remarks. I do not have any clue as to the date of the piece but I do remember it well because the entire design process was clearly an important part of NASA's beginnings and the privilege of having been connected with NACA and NASA through those most dynamic years is precious to me.[3]

FIGURE 4-2.

A NASA Distinguished Service Medal of the type awarded to astronaut Alan Shepard in 1961. (NASA)

---

2   "Lackluster Medals," *Time* magazine, 19 May 1961, p. 84.

3   "Ames Research Center View of NASA Logo Design," Letter from Harry J. DeVoto to Steve Garber of NASA History Office, 6 May 2001. NASA Headquarters Historical Records Collection, file 4542.

Applications of the Logos

The bold achievements of NASA during its embryonic era were spectacular and fundamental to the advancement of aerospace technology. The epic Moon-landing missions were achieved ahead of the schedule set by President Kennedy, sophisticated technologies and facilities were created to enable follow-on phases of terrestrial and universal exploration, and advanced aeronautical technologies for civil and military aircraft were developed.

The importance of NASA's programs—especially the critical Apollo Program—was not lost on the public, which brimmed with national pride inspired by the Agency and the symbolism of the NASA logos. The insignia, in particular, served its mission well as a centerpiece during both formal and informal media events ranging from the introduction of astronauts to press conferences and the visits of important stakeholders.

It appeared on NASA buildings and equipment, the flight suits of pilots and astronauts, research aircraft and facilities, and correspondence and reports. Visitors to the gift shops at NASA installations searched and scrambled for any souvenirs that displayed the insignia. Exposure of the iconic symbol became more widespread as the nation met the challenges of the space race, which culminated in lunar missions that took Modarelli's creation to the Moon.

AERONAUTICS

NASA's research aircraft and flying test beds displayed the insignia on various components of the airframe, typically the fuselage or wing. In many cases, the NACA gold band on the vertical tails of research aircraft was retained—with the NACA wings logo replaced by the letters "NASA" in black. This arrangement would be replicated on many NASA aircraft well into the 1990s.

FIGURE 4-3.

Majestic view of colorful NASA F-104N aircraft of the NASA Flight Research Center, in formation near Edwards Air Force Base in 1963. The tail bands are yellow, rather than gold. The NASA insignia is barely visible on the engine air intakes on the side of the fuselage. (NASA EC63-00221)

FIGURE 4-4.

NASA Langley began flight tests of its B737 research transport in 1973. In addition to the NASA insignia on its nose, the aircraft exhibited the insignia of the Federal Aviation Administration near its entrance door. The tail band was gold with stylized NASA letters in black. (NASA GPN-2000-001905)

FIGURE 4-5.

Showing the NASA insignia on its nose and a gold tail band, the XB-70A Ship 1 takes off on a research mission. The aircraft was flown by the NASA Flight Research Center in the mid-1960s for high-speed studies and for sonic-boom investigations. (NASA ED97-44244-2)

FIGURE 4-6.

NASA test pilot Pete Knight poses with the legendary X-15A-2 hypersonic research aircraft in 1965. A variation of the gold NASA tail band appears on the vertical tail, and the NASA insignia on the nose is backed by a gold flash. (NASA ECN-1025)

FIGURE 4-7.

Long before the Space Shuttle, NASA explored the flight behavior of advanced space vehicles known as lifting bodies. Here Milt Thompson, Chief Test Pilot of the NASA Flight Research Center, poses with a low-cost, unpowered test vehicle known as the M2-F1 in 1963. (NASA EC63-206)

FIGURE 4-8.

Rocket-powered lifting bodies were photographed at the Flight Research Center in 1972. Each carries the NASA insignia and gold tail band (L–R): The Air Force X-24A, the NASA M2-F2, and the NASA HL-10. (NASA EC69-2353)

FIGURE 4-9.

Low-speed research vehicles also carried the insignia. The insignia is displayed on the wing of the Boeing/Vertol VZ-2 tilt-wing Vertical Takeoff and Landing (VTOL) research aircraft shown at Langley in 1960. (NASA GPN-2000-001732)

Human Spaceflight

By far the most exposure for the NASA logos has been generated by the NASA human spaceflight program—the suborbital flights of Project Mercury, the challenging docking maneuvers of Project Gemini, the Moon landings of the Apollo Program, the launch of the Space Shuttle Program, and the evolution of the International Space Station. Whether worn as a patch on an astronaut's suit, affixed to boosters and space vehicles, or used as a backdrop during press conferences, the emblem is an integral part of NASA's program. The photographs that follow show the presence of the insignia during highlights of the human spaceflight program from this era.

FIGURE 4-10.

The first two groups of astronauts selected by NASA pose in 1963 behind NASA insignia. The original seven Mercury astronauts, selected in April 1959, are seated (L–R): L. Gordon Cooper, Jr.; Virgil I. Grissom; M. Scott Carpenter; Walter M. Schirra, Jr.; John H. Glenn, Jr.; Alan B. Shepard, Jr.; and Donald K. Slayton. The second group of NASA astronauts, named in September 1962, is standing (L–R): Edward H. White, II; James A. McDivitt; John W. Young; Elliot M. See, Jr.; Charles Conrad, Jr.; Frank Borman; Neil A. Armstrong; Thomas P. Stafford; and James A. Lovell, Jr. (NASA S63-01419)

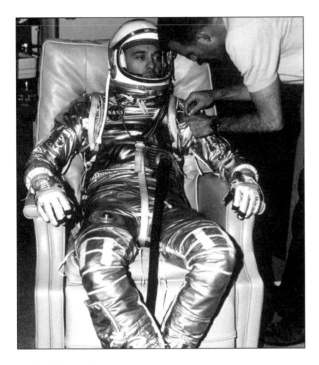

FIGURE 4-11.

One of the most popular Mercury astronauts was Alan Shepard, seen here in his spacesuit prior to the first U.S. piloted suborbital mission in 1961. His suit carries the NASA insignia, a tradition that became part of every piloted mission. (NASA MSFC-9248359)

FIGURE 4-12.

John Glenn smiles after becoming the first American to orbit Earth in his Mercury spacecraft Friendship 7 on 12 February 1962. (NASA MA6-39)

FIGURE 4-13.

President Kennedy honors John Glenn at Cape Canaveral on 23 February 1962. James Modarelli's NASA insignia is proudly displayed at the festivities. (NASA KSC-62PC-0014)

FIGURE 4-14.

Astronaut L. Gordon Cooper, Jr., prime pilot for the Mercury-Atlas 9 (MA-9) mission in 1963, relaxes while waiting for weight and balance tests to begin. The NASA insignia was worn in the same location on flight suits during Project Mercury. (NASA S63-03974)

FIGURE 4-15.

Before it could embark on a Moon mission, NASA had to demonstrate the feasibility of a rendezvous mission with an orbiting target vehicle in the Gemini Program. NASA insignia and mission patches are worn by Astronauts Walter M. Schirra, Jr. (seated), command pilot, and Thomas P. Stafford, pilot, Gemini 6 prime crew, as they go through suiting up exercises in preparation for their Gemini 6 flight on 20 October 1965. (NASA GPN-2000-001478)

FIGURE 4-16.

The NASA insignia dominates the side of the Project Gemini Control Center building at Cape Canaveral in 1964. (NASA KSC-64C-2699)

FIGURE 4-17.

The astronauts of Apollo 11 (L–R): Neil Armstrong, Michael Collins, and Buzz Aldrin pose in spacesuits before their historic flight. The NASA insignia and American flag adorned their suits. (NASA GPN-2000-0011644)

FIGURE 4-18.

Insignia on the Moon.
Astronaut Buzz Aldrin
steps down from the LEM
in preparation for a task.
The insignia carried on
the front of the crew's
suits was hidden by
equipment, but the NASA
insignia on their back-
packs can be seen here.
(NASA AS11-40-5869)

NASA TECHNICAL NOTE   NASA TN D-5409

NASA TN D-5409

ANALYSIS OF THE FLAT-SPIN CHARACTERISTICS OF A TWIN-JET SWEPT-WING FIGHTER AIRPLANE

by Joseph R. Chambers, James S. Bowman, Jr., and Ernie L. Anglin

Langley Research Center
Langley Station, Hampton, Va.

NATIONAL AERONAUTICS AND SPACE ADMINISTRATION • WASHINGTON, D. C. • SEPTEMBER 1969

FIGURE 4-19.

The covers of NASA technical reports published from 1959 to 1975, such as this technical note, included the NASA seal.

PUBLICATIONS

NASA publications during this period—for example, technical notes—were published with an image of the NASA seal on the cover. These documents were subjected to a thorough in-house peer review before being considered for publication and release to the industry, military, universities, and the public. Such publications were of high quality and widely used by the aerospace community.

CHAPTER 5

Controversy, 1975–1991

The NASA Logotype

The NASA insignia served the Agency well for 16 years, becoming an iconic centerpiece during the excitement and national pride of the Apollo years. In 1975, however, NASA Headquarters decided that the insignia for the Agency should take on a more contemporary appearance. James Modarelli's original logo, which had always been referred to as "the insignia," was first dubbed "the meatball" in 1975 by Frank "Red" Rowsome, head of technical publications at NASA Headquarters, to differentiate it from a new logo called the NASA logotype.[1] In 1974, as part of the Federal Graphics Improvement Program of the National Endowment for the Arts, NASA replaced the complex meatball with a stripped-down, modernist interpretation in which even the cross stroke of the As was removed. The logotype was designed by Danne

and Blackburn, a New York firm that specializes in corporate identity. Bruce N. Blackburn of the firm designed the American bicentennial logo, which was painted on the NASA Vehicle Assembly Building at Kennedy Space Center in 1976.

The Federal Graphics Improvement Program, active from 1972 to 1981, was part of the Federal Design Improvement Program. A panel of prestigious graphic designers convened to critique the graphics of participating federal agencies. This critique was not limited to paper products; planes, ambulances, and even the Congressional Record were cited for redesign. Ultimately, more than 45 government agencies revamped their graphics under this mandate.[2]

In addition to providing a more contemporary look, the logotype was in part justified by difficulties encountered with reproducing the colors of the meatball in the printing process before the age of

1   "Employees Rededicate Famous NASA Symbol." *The Lewis News*, in-house newsletter, Volume 34, Issue 11, November 1997, p. 1.

2   See *http://arts.gov/article/setting-standard-nea-initiates-federal-design-improvement-program* (accessed 5 November 2014).

NASA

FIGURE 5-1.

The NASA logotype replaced the insignia in 1975. (NASA C-1997-4062)

FIGURE 5-2.

The American bicentennial logo was designed by Bruce N. Blackburn. It was added to the Vehicle Assembly Building at the Kennedy Space Center in 1976 and replaced by the meatball in 1998.

high-quality digital printers. In addition, recognition of some of the elements—for example, the stars—in the meatball design was difficult for some viewing conditions. The color of the new insignia was NASA Pantone 179 (red). When compared to the red wing in the meatball insignia, the logotype color was more of an orange-red. Against a white background, the logotype was shown in red, black, or a warm gray.

Acceptance of the new logotype insignia was far from universal within the Agency. Many longtime NASA employees were dismayed by the replacement of their beloved meatball with the trendy new insignia, which they called the "worm" in a truly derogatory sense. The new insignia was never fully endorsed by the NASA old-timers, leading to a raging controversy. Nonetheless, the NASA logotype was adopted and would serve as the Agency's insignia for the next 17 years. During that period, NASA would continue to make unprecedented scientific advancements, but the occurrence of the Challenger disaster, the resulting termination of Shuttle flights, and other Agency project problems resulted in a perceived downturn in morale, especially at the Centers.

The New Insignia

The adoption of the new NASA logotype as a new insignia at the field Centers was received with surprise and mixed emotions. The tone of unrest was exemplified by the announcement of the unanticipated insignia in Langley's *Researcher News*, in-house newsletter,

A new graphics design program is being implemented for NASA. The program is planned to improve the agency's ability to communicate, provide uniformity for all NASA graphics, and save time and money. The new program, called a graphics system, will cover all means of communi-cation, both inside and outside NASA: stationery, technical and management publications, vehicles and aircraft, buildings and direction signs, news releases, and a new Logotype (symbol).

The new symbol is the most obvious—and the most controversial—part of the graphics system. The symbol will replace the NASA "Meatball" for all agency identification. The NASA seal will be retained, however, for "formal" uses such as service pins, award certif-icates, speakers lecterns, the NASA flag, and for other occasions.

The new symbol, reproduced here, incorpo-rates the letters of the NASA acronym, with let-ters stylized and reduced to their simplest forms. The single width of the letters is designed to give a feeling of unity and technological precision to the symbol. The curves of the letters represent fluidity and continuity from a design point of view. The "As" are used without their cross-strokes to give a feeling of vertical thrust.

The words "National Aeronautics and Space Administration" will accompany almost every use of the new symbol. According to designers, millions of people know about NASA but many of them don't know what the acronym means.

The new graphics system has been accepted by NASA Administrator Dr. James C. Fletcher. At the time he stated:

"We at NASA believe that design excellence is not a luxury, but rather a necessity. We believe that it is important to constantly upgrade the quality of our graphics in order to improve communications with the citizens of our coun-try and to develop a closer relationship between graphic design and program operation so that design becomes a tool in achieving the program objectives of NASA."

FIGURE 5-3.

This humorous, tongue-in-cheek "Goodbye Meatball" cartoon captures the abrupt-ness of the worm insignia replacing the meatball insignia in 1975. (Contributed by James H. Cato)

To keep changeover costs at a minimum, many elements will be gradually changed, including such items as vehicles and signs. The complete changeover period is expected to take several years.

The new graphics system began as part of the Federal Design Improvement Program, imple-mented by the National Endowment for the Arts, which began in 1972. NASA's participation began when a panel of graphic design experts recommended several ways in which the Agency could improve its ability to communicate.[3] The New York design team of Richard Danne and Bruce Blackburn developed at the NASA graphics system and will help implement it. The firm also designed the National Bicentennial symbol.[4]

In reality, Administrator Fletcher and his Deputy Administrator, Dr. George Low, had misgivings about the new design.[5] Both Fletcher and Low expressed their uncomfortable positions to the designers during their first design presentation, but the new logo was nonetheless approved and implemented.[6]

IMPLEMENTATION

Robert Schulman of NASA Headquarters was chosen to be NASA's monitor of the redesign effort as it continued to evolve over the next decade. The Standards Manual for the use of the logotype ultimately consisted of over 90 pages, covering applications to ground vehicles, aircraft, the Space Shuttle, signs, uniform patches, public service film titles, space vehicles, and satellite markings.

Co-designer Richard Danne recalled that, in his opinion, the difficulty encountered regarding acceptance of the new logo was caused by how the Agency decided to alert the Centers of the change,

> The Agency made the decision to alert the various Centers to the new Program by sending Executive stationery to each Center Director. That Stationery displayed the new NASA Logotype and it would be the first time they were informed of the graphics program and image change…. Centers were like fiefdoms, they enjoyed their freedom and their provincial specialties. At the time of our redesign, there was resistance to almost anything emanating from Headquarters…. Those letterhead "gifts" coming out of Washington and sporting the new Logotype proceeded to detonate across the country… and all hell broke loose![7]

The managers at the NASA Centers were upset over the unexpected announcement, and the employees were even more inflamed. The perception of the program advocates was that the younger NASA employees preferred the new graphics program, but the older employees were, in general, upset over the change. There appeared to be no middle ground for the situation; employees and NASA's partners either enthusiastically endorsed the logotype or despised it.

In 1984, the NASA logotype was awarded the prestigious Award of Design Excellence of The Presidential Design Awards by President Reagan in Washington. Despite this recognition, the schism in acceptance of the logotype at the Centers persisted throughout its existence. To the relief of many, it was replaced by Modarelli's meatball design following the appointment of Daniel S. Goldin as the new administrator of NASA by President George H. W. Bush on 1 April

---

3   Panel members included Marion Swannie, IBM Design Program Manager; Alvin Eisenman, Chairman of Yale University's Graphics Department; Paul Rand, nationally known graphics designer; Howard Paine of National Geographic magazine; and John Leslie of the U.S. Department of Labor.

4   *The Researcher News*, Langley's in-house newsletter, Volume 14, Issue 4, 21 February 1975, p.1.

5   Hilary Greenbaum, "Who Made Those NASA Logos?" *The Sixth Floor*, 3 August 2011. *http://6thfloor.blogs.nytimes. com/2011/08/03/who-made-those-nasa-logos/?_php=true&_ type=blogs&pagewanted=p&_r=0* (accessed 31 August 2014).

6   A detailed discussion by designer Richard Danne regarding his perspective on the design and advocacy problems for the NASA logotype is available online at *http://www. thisisdisplay.org/features/the_nasa_design_program* (accessed 1 September 2014).

7   Ibid.

1992. The next chapter will discuss why Goldin made this change.

The Shuttle Emblems

The emergence of the Space Shuttle during the 1970s resulted in a new emblem for NASA's Space Transportation System (STS) on 18 February 1977. Known as the Space Transportation System Program Badge, the emblem was triangular, pointed upward, and had a stylized drawing of the Space Shuttle in launch configuration. Its colors were a white and grey shadow-edged Shuttle over a background of two shades of blue, and a gold and white edge. "SPACE SHUTTLE" was inscribed horizontally at the base in white letters.[8]

The STS badge was designed by the NASA Johnson Space Center, which also designed "Add-On Bars" to augment the badge and provide subordinate mission-related identity. The bars were rectangular, gold and white edged, with one or two lines of text on a red band. They were positioned horizontally at the base of the badge. Over 15 different identifiers were made with titles such as Orbital Flight Test, Approach and Landing Test, Johnson Space Center, and Space Shuttle Main Engine.[9]

Early in the Shuttle program, NASA management planned to use the add-on bars as numeric indicators of individual missions. The astronauts, however, overruled this approach and instead initiated individual mission patches that became an integral part of the human spaceflight program. NASA formally referred to mission patches as the NASA Astronaut Badge.

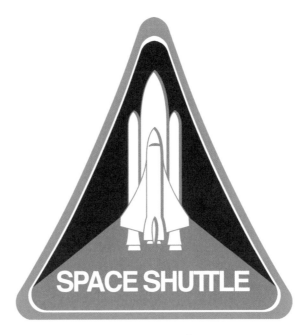

8   Federal Register, Volume 52, Number 231, 2 December 1987, p. 45818.

9   *http://www.collectspace.com/ubb/Forum18/HTML/000420.html* (accessed 15 September 2014).

FIGURE 5-4.

The NASA Space Transportation System Program Badge was established in February 1977.

Shuttle crews prior to the mid-1980s typically wore the STS badge along with the NASA logotype insignia during preflight and mission photo opportunities. After the Challenger disaster, the STS badge was rarely seen on uniforms and flight suits. From the 1990s on, Shuttle crews wore uniforms with their mission insignia and the NASA meatball insignia.

Applications of the Logotype

As expected, the changeover from the meatball to the logotype insignia was slow due to the cost and labor required. Some of the changes—especially those involving building identifiers—occurred relatively quickly, but other more costly changes took years because the NASA meatballs had been extensively displayed throughout the field Centers. The process was also significantly delayed by the reluctance of the old-timers to dismiss their favorite insignia.

BUILDINGS

The introduction of the new NASA logotype insignia in 1975 required NASA installations to identify buildings with the new insignia. The changeover in logos for large buildings such as aircraft hangars prompted different actions at the various NASA Centers. For example, the flight hangars at the Ames-Dryden Flight Research Facility immediately complied with the new directive, whereas Langley never removed its meatball from the hangar and only added the logotype in late 1991.

The Unitary Plan Wind Tunnel complex at the Ames Research Center has long been a key national facility supporting the U.S. civil and military aerospace communities. Thousands of tests of almost every commercial and military aircraft have been conducted there since it began operations in 1956, and testing continues today. The facility was designed to have three test sections using only one set of drive motors. A turning vane/valve arrangement is used to divert the airflow into one test section while the other two are readied for another test or undergo maintenance. The meatball insignia had been painted in a prominent position on the turning vane/valve used to separate flow between test sections.[10] This application was but one of many in which the old meatball was given preference over the new worm insignia.

---

10 *The Astrogram*, in-house Ames newsletter, Volume XXXII, Number 25, 31 August 1990. A historical preservation study of the facility is available online at *http://historicproperties. arc.nasa.gov/unitary.html* (accessed 3 September 2014).

FIGURE 5-5.

The NASA logotype insignia is exhibited over the entrance to the main building at the NASA Ames-Dryden Flight Research Facility in 1991. In this photograph the logo appears behind the X-1E research aircraft with its NACA tail band. (EC-91-485-1)

FIGURE 5-6.

The research aircraft of the NASA Ames-Dryden Flight Research Facility were photographed in 1985 in front of flight hangars marked with the NASA logotype insignia. The Shuttle hangar is in the background at the upper right. (EC-91-485-1)

FIGURE 5-7.

The Ames Unitary Plan Wind Tunnel was declared a National Historical Landmark in 1985. The flow diverter section is proudly adorned with the meatball. The plaques on display in front of the tunnel discuss the contributions of the facility. (NASA ACD06-0213-001)

AIRCRAFT

During the period 1975–1991, NASA research aircraft appeared with a variety of insignia and, for the first time, an increased number of logos that identified partners and sponsors for research projects. Many of the paint schemes were chosen to enhance the visibility of the vehicle as well as airflow over and around its surfaces.

FIGURE 5-8.

The NASA F-18 High Alpha Research Vehicle (HARV) had a special black finish to enhance visual analysis of flow characteristics over the airplane at high angles of attack. The NASA logotype insignia and the aircraft's NASA identification number were therefore painted in a nonstandard white color. (NASA EC92-10231-2)

FIGURE 5-9.

The NASA logotype insignia was displayed on the vertical tail of the X-29 research aircraft in 1990. The vehicle also displayed the names of NASA's partners including Grumman, the Defense Advanced Research Projects Agency (DARPA), and the U.S. Air Force. (NASA EC92-10231-2)

FIGURE 5-10.

The Highly Maneuverable Aircraft Technology (HiMAT) remotely piloted vehicle exhibited the NASA logotype insignia during its research program at the Ames-Dryden Flight Research Facility in 1980. (NASA EC-14281)

FIGURE 5-11.

As might be expected, during the decade following the introduction of the NASA logotype insignia, aircraft markings at various NASA facilities involved a mixture of old and new logos. Observers were often confused by scenes of non-uniformity in NASA markings in photographs, at airshows, and during other public events. Here, a formation flight of Dryden F-15 and F-104 aircraft in 1980 shows the differences in NASA markings commonly encountered at the time. (NASA EC80-14126)

FIGURE 5-12.

During the age of the logotype, several aircraft retained the meatball insignia. Here NASA's F-106B research aircraft performs a flight above NASA Langley Research Center in 1979. The aircraft was used by Langley for storm hazard (lightning strike) and vortex-flap flight research studies. (NASA L79-7204.1)

SPACE SHUTTLE

In 1977, NASA began a yearlong study of the aerodynamic and structural characteristics of the Space Shuttle Enterprise and the B747 Shuttle Carrier Aircraft during mated flights at Dryden. The combination conducted five free-flight tests to evaluate the unpowered gliding and landing handling qualities of the orbiter prototype.

Enterprise and Challenger always had the worm insignia on their wings. Beginning in 1998, the insignia for Endeavour, Discovery, Columbia, and Atlantis was changed from the worm to the meatball.

Following the initial flights at Dryden, the Space Shuttle Columbia (which also bore the NASA worm logo at the time) successfully rocketed into space for the first Shuttle mission on 12 April 1981.

Space Shuttle operations became routine, with countless successful and productive scientific and space exploration missions being conducted. The entire nation became accustomed to NASA's remarkable success—until 28 January 1986 and the Shuttle Challenger disaster.

Shuttle Discovery was launched on a highly successful Return-to-Flight mission on 29 September 1988, which inspired a feeling of optimism at the Agency and throughout the nation for the human space program. The NASA logotype and the meatball were both displayed on the astronauts' suits, and Discovery carried the logotype insignia on its rear fuselage for the mission. The mission patch, which included a rendition of the red wing of the meatball, emphasized the successes of the previous NASA and the beginnings of a

FIGURE 5-13.

The Space Shuttle Enterprise and Shuttle Carrier Aircraft display the NASA logotype insignia during a test flight in February 1977. Note the gray color of the insignia on the orbiter. Enterprise never flew in space. (NASA ECN-6882)

new NASA. The themes of the STS-26 insignia were: a new beginning (sunrise), a safe mission (stylized launch and plume), the traditional strengths of NASA (the red wing of the meatball), and a remembrance of the seven astronauts who died aboard Challenger (the

seven-starred Big Dipper). The insignia was designed by artist Stephen R. Hustvedt of Annapolis, Maryland.[11]

During this phase of the Shuttle Program, astronauts typically wore flight suits with patches of both the logotype insignia and the meatball. The logotype was worn until March 1996 for the flight of STS-76, after which its use was discontinued.

11 See *http://www.jsc.nasa.gov/history/shuttle_pk/pk/Flight_026_STS-026R_Press_Kit.pdf* (accessed 10 November 2014).

FIGURE 5-14.

Enterprise mounted in launch configuration during a fit check at Vandenberg AFB in 1985. The orbiter shows its logotype insignia on the fuselage and right wing. (Air Force Photo DF-ST-99-04905)

FIGURE 5-15.

Astronauts John W. Young, left, crew commander, and Robert L. Crippen, pilot for the STS-1 mission, pose with a Shuttle model in 1979. They wear patches of the logotype insignia and stand before a NASA flag with the NASA seal. The model also exhibits the logotype. (NASA KSC-79PC-0271)

FIGURE 5-16.

President Ronald Reagan (in tan suit) joined a huge crowd in welcoming the Shuttle Columbia following its landing at Edwards at the end of the fourth Shuttle mission on 4 July 1982. The NASA logotype insignia was prominently displayed on the prototype Shuttle Enterprise, hangars, and personal attire worn by onlookers. (NASA EC82-18992)

FIGURE 5-17.

The crew of the Shuttle Discovery before their STS-26 Return-to-Flight mission in 1988. Crew members were: Back row (L–R): mission specialists John M. "Mike" Lounge, David C. Hilmers, and George D. "Pinky" Nelson. Front row (L–R): Richard O. Covey, pilot, and Frederick H. "Rick" Hauck, commander. (NASA GPN-2000-001174)

FIGURE 5-18.

Shuttle Discovery lands at Edwards after completion of the STS-26 Return-to-Flight mission on 3 October 1988. Like all earlier Shuttles, Discovery displays the logotype insignia. (NASA GPN-2000-001174)

FIGURE 5-19.

Shuttle Atlantis blasts off from Kennedy Space Center on STS-34 to launch the Jupiter deep space exploration spacecraft Galileo on 18 October 1989. Atlantis carried the NASA logotype insignia on its right wing. The Galileo spacecraft provided scientists with a wealth of data on the largest planet in our solar system and its numerous moons, until 2003 when the spacecraft was deorbited. (NASA)

SATELLITES

NASA launched the Hubble Space Telescope into low-Earth orbit from the Space Shuttle Discovery in 1990, enabling astronomers to observe details of the universe to an unprecedented degree. Hubble continues to provide incredible views of space and time and is one of the highlights of NASA's accomplishments. Several successful Shuttle servicing missions to enhance the satellite's performance have followed.

FIGURE 5-20.

The Hubble Space Telescope is seen in Shuttle Columbia's payload bay following successful repairs and reconfiguration during the STS-109 mission in March 2002. The telescope still exhibited the logotype insignia from its insertion into orbit in 1990. (NASA STS109-318-208)

PUBLICATIONS AND CORRESPONDENCE

From 1975 to 1991, the technical note continued to be NASA's pre-ferred document for the dissemination of technical information. The NASA seal remained on the covers of the notes, and during the bicen-tennial year of 1976, the documents also carried the bicentennial logo.

NASA TECHNICAL NOTE NASA TN D-8260

NASA TN D-8260

A DESIGN APPROACH AND SELECTED WIND-TUNNEL RESULTS AT HIGH SUBSONIC SPEEDS FOR WING-TIP MOUNTED WINGLETS

Richard T. Whitcomb

Langley Research Center
Hampton, Va. 23665

NATIONAL AERONAUTICS AND SPACE ADMINISTRATION · WASHINGTON, D. C. · JULY 1976

FIGURE 5-21.

NASA Technical Memorandum 101701

Effects of Wing Sweep on In-Flight Boundary-Layer Transition for a Laminar Flow Wing at Mach Numbers From 0.60 to 0.79

Bianca Trujillo Anderson and Robert R. Meyer, Jr.
Ames Research Center, Dryden Flight Research Facility, Edwards, California

1990

NASA
National Aeronautics and
Space Administration
Ames Research Center
Dryden Flight Research Facility
Edwards, California 93523-0273

FIGURE 5-22.

During the 1976 celebration of the nation's bicentennial, NASA technical reports were printed with the NASA seal and the bicentennial logo on the cover. (NASA)

Example of the cover of a NASA technical report typical of the early 1990s with the logotype insignia on display. (NASA)

Anniversary Logos

In 1990, NASA celebrated the 75th anniversary (1915–1990) of the combined operations of the NACA and NASA. Ceremonies were held to observe the event, and souvenir medallions were minted and distributed to employees, along with special logo stickers. The medallions featured a NASA side with a simplified version of the NASA meatball and an NACA side with a rendition of the NACA seal.

FIGURE 5-23.

A medallion was issued to all NASA employees in 1990 on the occasion of the 75th anniversary of the NACA and NASA. (Joseph Chambers)

FIGURE 5-24.

Stickers and pins of a special anniversary logo were also distributed to employees. (Contributed by Peter Jacobs)

CHAPTER 6

Back to the Future, 1992–Today

Rebirth of the Meatball

New NASA Administrator Daniel S. Goldin arrived via NASA airplane for his first tour of the Langley Research Center on 22 May 1992. As the plane taxied onto the hangar apron, a passenger looked at the NASA logotype insignia on the hangar and asked, "Why in the world do we have that awful logo?"[1]

Immediately after he arrived, Goldin asked his special assistant George W. S. Abbey and Langley Center Director Paul F. Holloway what he could do to improve what he perceived to be the sagging morale of employees at the NASA Centers. Abbey pointed to the meatball insignia that had been retained on the Langley hangar and said, "Restore the Meatball!"[2]

Holloway showed Goldin his personal business card, which still carried the meatball insignia and told him that he (Holloway) wanted to work for that Agency. He also cited the ridiculous procedure he had gone through to legally carry the card. The requirement for use of the old meatball emblem was to have a letter on file signed by the Administrator. The same requirement existed for astronauts to use it on spacesuits. All exceptions had to be approved by the Administrator.[3]

Later that day Goldin addressed the Langley employees and the entire NASA organization via NASA TV,

> Now let me tell you, we have a little surprise today. In honor of that spirit, it seems only fitting that the original NASA insignia—affectionately we know it as the "Meatball"—will be part of

1   Paul F. Holloway, telephone interview by Joseph Chambers, 15 July 2013.

2   Paul F. Holloway, interview by Joseph Chambers, 15 July 2013; Kirk Seville, "Worm Turns," *Newport News Daily Press*, 23 May 1992. Available online at *http://articles.sun-sentinel.com/1992-05-23/news/9202090944_1_stationery-and-repaint-goldin-worm* (accessed 31 August 2014), and

http://www.wired.com/2014/09/weird-facts-about-5-of-the-worlds-most-famous-logos/?mbid=social_twitter (accessed 4 September 2014)

3   E-mail communication with Armstrong Flight Research Center public affairs specialist Cam Martin, 9 January 2013.

FIGURE 6-1.

The Langley flight hangar displayed both the NASA logotype and the meatball insignia during Administrator Goldin's visit in 1992. Although taken 2 years later, the photo shows the markings at the time of the visit. (NASA 1994-L-04816)

our future. I know you feel this way too, because large numbers of you have stopped and told me so as I've been visiting all the Centers around the country… This does not mean that as of today we will throw out all the stationary and repaint every building and vehicle—although we'd love to do it… Over time, the NASA symbol of old, the affectionate Meatball, will replace the slick NASA logo and slowly it will die into the horizon and never be seen again… Take pride in the symbol that stood for NASA excellence in the past—and now—looks to the world-class NASA

of today and tomorrow… The can-do spirit of the past is alive and well. The magic is back… The old NASA insignia is back because the men and women of NASA ordered it to be back. A classic, it never goes out of style![4]

Goldin had arrived on a Thursday, and by Friday morning NASA's worm was officially gone, and the meatball was back. The Administrator wore a borrowed meatball pin during his visit, and a meatball sign was hastily affixed to the podium during his presentation to the employees and media.[5]

Many of the "old-time" managers at Langley—including Holloway—had been reluctant to follow the Headquarters directives for implementing the new logotype emblem and, as a result, the Langley hangar had not displayed the new insignia from 1975 until late 1991. Instead, the hangar retained the meatball insignia painted on the facility's "tail door." As an additional affront to the worm, in 1973 the Center had received a Boeing B737 transport that it subsequently converted into one of the world's most productive research airplanes. When the airplane was scheduled for painting with NASA livery after arrival, the flight operations organization proposed putting a meatball insignia on the nose, despite the existing Agency directive to remove meatballs from aircraft markings. Langley management turned a blind eye to the directive; the

4   "Administrator Goldin Brings 'New NASA' Message to Langley," *The Researcher News,* in-house Langley newsletter, Volume 6 Issue 9, 5 June 1992, p. 1.

5   Goldin borrowed the meatball pin from Cam Martin of the Office of Public Affairs at Langley. At the time, the pins were "illegal" and no longer sold in Langley gift shops. Martin had bought the pin years before and hurried to his nearby home to retrieve it in time for Goldin's briefing. Communication with Cam Martin, 9 January 2013.

FIGURE 6-2.

A meatball sign was hurriedly found and affixed to the podium above the soon-to-be removed NASA logotype during Administrator Goldin's briefing. (NASA 1992-L-06195)

FIGURE 6-3.

Goldin expresses his affection for the meatball insignia on the nose of the Langley B737 research transport. (NASA 1992-L-06225)

meatball was placed on the nose and the NASA logotype was painted on the tail.

In anticipation of Administrator Goldin's first visit to Langley, priority had been given to creating a highly visible NASA logotype on the hangar building. A great effort was expended to make sure that the Center was in compliance with the logotype directives. As previously discussed, however, the visiting entourage's perception of the logotype insignia upon arrival, and Goldin's ensuing discussions with George Abbey and Paul Holloway, led to the demise of the worm. Later in the day, Goldin posed for a photograph in the Langley aircraft hangar; in the photograph, he extended his arms around the NASA meatball on the nose of the B737 in an admiring fashion. After his departure, a scramble ensued for many years to hide the worm at all NASA Centers in compliance with Goldin's decision.[6] To the amusement of many, the situation evolved into a NASA-wide "worm hunt."

Although the old-timers were jubilant over the return of their beloved meatball insignia, Goldin received strong criticism and requests for reconsideration of his decision from some of the media and the Endowment for the Arts.[7]

In his comments to employees, Goldin had directed them not to waste money by throwing out

6   E-mail communications with Cam Martin, 9 January 2013 and 5 September 2014.

7   Kathy Sawyer, "Logo Change Sends NEA Into Orbit: Federal Design Director Urges NASA Chief to Reassess Decision," *Washington Post,* 15 June 1992. Files of the Institute of Heraldry, Department of the Army, Fort Belvoir, VA, to be deposited in HRC. One particularly critical position was taken by the online site *NASA Watch.* For example, see "Small Wonder People Under Stress at NASA When You Behave This Way" at *http://nasawatch. com/archives/1999/12/,* 26 October 1999 (accessed 8 September 2014). The criticism continued past the turn of the century. For example, see "Art of the Seal" by Alice Rawsthorn, *New York Times Style* magazine, 5 March 2009.

worm-infested stationary or by repainting worm-covered space vehicles.[8] NASA employees took his comment to heart—Langley's hangar would retain the NASA logotype insignia until late 1999, and many research aircraft across the Agency would retain their logotype tail markings for several years thereafter. The Space Shuttles flew with the emblem well into the 1990s. Eliminating the worm significantly impacted costs and schedules. As will be discussed, the overlap of NASA insignias during the transition led to confusion among observers.

The "Wormball"

Following Administrator Goldin's decision to bring back the meatball, the magazine *Quest: The History of Spaceflight Quarterly* published articles on the turmoil between the supporters of the worm insignia and those of the meatball insignia. In the first article, a *Quest* artist created a composite logo in which he combined both the worm and meatball into one design, with the worm replacing the standard NASA letters of the meatball.[9] Calling the design "The best of both worlds?" was a humorous comment on the logo situation within NASA.

In its next issue, *Quest* reported that Modarelli had read the article and wrote, "It was interesting... since I submitted an almost identical design to NASA Headquarters shortly after the NASA worm logo was adopted in 1975. Naturally it was turned down...."

FIGURE 6-4.

The unofficial "wormball" was proposed as a "make everyone happy" design for the NASA insignia. (*Quest: The History of Spaceflight Quarterly*)

Interest in the "wormball" inspired a movement to have NASA adopt it as the logo. In response to the issue, NASA Headquarters quickly made it clear that the "wormball" would never be adopted.[10]

Overdue Recognition

By the 1960s, Modarelli had still not received proper recognition for his design of the meatball. This situation was brought to the attention of Abe Silverstein, Director of the Lewis Research Center, who pointed out the grievous oversight in a forceful letter to NASA Headquarters in 1967.

---

8   Ibid.

9   *Quest: The History of Spaceflight Quarterly*, Summer 1992, Volume 1, number 2, p. 29. The artist was Dan Gauthier. Others advocated the concept; see Keith Cowing, "New NASA Logo Designs," *NASA Watch*, 25 April 2010, *http://nasawatch.com/archives/2010/04/new-nasa-logo-d.html* (accessed 20 September 2014).

---

10   *Quest: The History of Spaceflight Quarterly*, Fall 1992, Volume 1, Number 3, p. 45.

In glancing over NPD 1020. lA, September 13, 1967, covering the NASA Official Seal, Insignia and Flags, I noticed that paragraph 5 thereof gives credit for the design of the Insignia to the Army Institute of Heraldry.

It had been my recollection that Jim Modarelli had been largely responsible for the design of the NASA Seal, Insignia, and Pins. Accordingly, I reviewed the history with Jim and verified my recollection. It is true that a number of groups had their fingers in the job, however, there is no question that, starting with the Seal and proceeding with the Insignia and the Pins, Jim Modarelli not only contributed the initial basic conceptual design, but was also responsible for the final refinements. Jim mentioned that after adoption of the Seal, Keith Glennan was dissatisfied with the submission of an industrial design firm that had been subsequently commissioned to develop the Insignia. He then selected a design that Jim worked up. The story of the NASA Pins seems to be more of the same. As far as the Army Institute of Heraldry is concerned, their contribution seems to have primarily been the selection of the ribbon design associated with the NASA medals.

Although the matter could be considered trivial, I feel that an injustice has been done in failing to give credit to Jim Modarelli whom I am satisfied was the principal contributor to the design of the NASA Seal, insignia, and Pins. I suggest that the matter be reviewed and appropriate action be taken to redress the slight. Should any background evidence be required, we can provide copies of many of the original and succeeding drawings, photos, art work, etc. that led to the final design and that were largely prepared here at Lewis.[11]

Former NASA Administrator Glennan offered another emotional endorsement of Modarelli's work in a letter to Administrator Goldin in June 1992, a few months after Goldin brought back the meatball.

Quite coincidentally, I became aware of your decision to re-instate the original NASA insignia and retire the one that was substituted for reasons I could never fathom. I will recall the day when several of our top staff met in my office to select the winning design from perhaps eight or ten that had been submitted. After giving everyone a chance for his apple, I said we were wasting time and I chose the one submitted by Jim Modarelli who was at the Lewis laboratory of the old NACA. I believe that he later designed the first NASA commemorative stamp recognizing the Mercury program. Former NASA people are now urging that Jim Modarelli be recognized for his early contributions to the NASA operation in the field of public information. And I strongly endorse these suggestions although I have no ideas as to the type of recognition which might be appropriate. On my bedroom wall I have hanging a large plaster replica of what I've always called the NASA seal and I'm very glad that it is once more the official NASA insignia.[12]

---

11 Letter from Lewis Director Abe Silverstein to Harold B. Finger, Associate Administrator of NASA Office of Organization and Management, 25 September 1967. NASA HRC File 61.

12 Letter from Dr. T. Keith Glennan to Daniel Goldin, Administrator of NASA, 23 June 1992. NASA Headquarters HRC File 4543.

FIGURE 6-5.

James Modarelli (left) and his wife Lois receive a special award from NASA Administrator Daniel Goldin in 1992. The award citation stated, "To the creator of NASA's original 'Meatball' insignia. In honor of your lasting contributions to the proud spirit of the National Aeronautics and Space Administration. 21 October 1992." (NASA Glenn 1992-09961)

Administrator Goldin made sure Modarelli finally received formal recognition for his contribution to the culture of the Agency. In October 1992, Goldin presented a well-deserved special award to Modarelli at the Lewis Research Center.

On 1 October 1997, 18 years after Modarelli's retirement, the NASA Lewis Research Center replaced its weatherworn 35-year-old meatball insignia on the Lewis hangar with a new 20-foot-diameter version. Nearly 250 Lewis employees gathered inside the hangar to rededicate the famous logo. Following the rededication ceremony, the Lewis employees were invited to sign their names on the back of the insignia before it was placed on the hangar. The honor of being the first to sign was, of course, given to Modarelli.[13] As guest of honor, Modarelli addressed the crowd and received an enthusiastic response to his remarks.

The Designer Passes

James Modarelli started his career at Lewis in 1949 as a technical equipment illustrator and was named division chief of the Management Services Division in 1954. He left Lewis and served as exhibits chief at NASA Headquarters for the years 1959–1961), then returned to Lewis where he served as chief of the Technical Publications Division before retiring in 1979 as chief of the Management Services Division after a 30-year career. He was particularly popular with his employees—a member of his division said, "He is one of only a few people I've ever known who has never had a negative comment made about them." Modarelli was socially active at Lewis and in community affairs and was an avid participant in jogging and

13 "Employees Rededicate Famous NASA Symbol," *The Lewis News* Volume 34 Issue 11, November 1997, p. 1.

FIGURE 6-6.

The new 20-foot-diameter meatball insignia for the Lewis hangar is prepared for installation in 1997. (NASA C-1997-4062)

FIGURE 6-7.

James Modarelli addresses the attendees at the rededication ceremony for the meatball insignia at the Lewis hangar. (NASA C-1997-3940)

racing contests.[14] He was part of a summer camp that offered pre-apprentice training classes to 102 minority high school graduates, he assumed a leadership role in the NACA triennial inspections, and he helped organize VIP retirement parties and a variety of employee morale-building activities.[15] Modarelli said he looked back with pride on his years at NASA with a remarkable team of dedicated people.

Modarelli had, of course, been very disappointed when his meatball design was replaced by the worm. He said, "I didn't have any objection to a new logo for NASA, but I have always felt that the 'Worm' didn't complement the NASA seal and was too flashy and trendy. Since the original insignia ('Meatball') is made up of the same elements as the seal, the two complement each other. There is a definite relationship."[16]

James Modarelli died on 27 September 2002. He had seen his meatball design flourish as the prime insignia of the Agency and become instantly recognized the world over—it had even visited the Moon. His highly artistic concept lives on today as a perpetual reminder of the pride, dedication, and accomplishments of NASA.

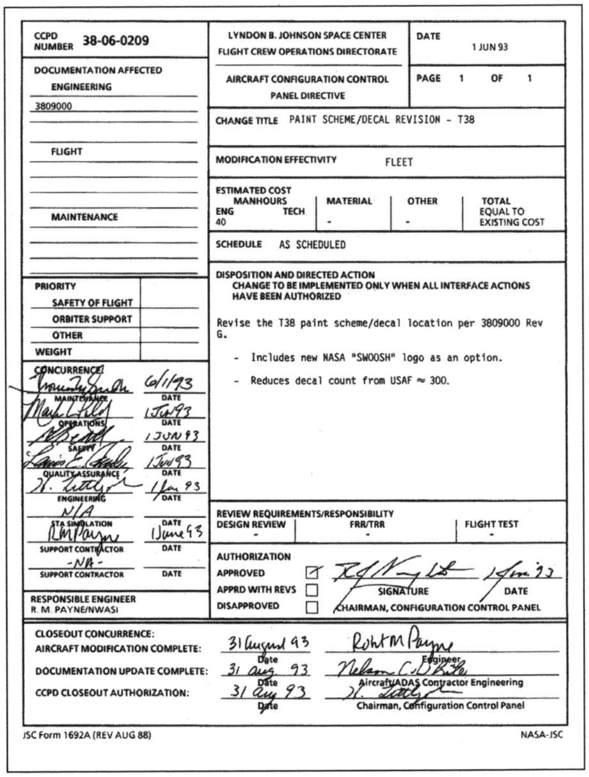

CCPD NUMBER 38-06-0209
LYNDON B. JOHNSON SPACE CENTER
FLIGHT CREW OPERATIONS DIRECTORATE
DATE 1 JUN 93
DOCUMENTATION AFFECTED
ENGINEERING
3809000
FLIGHT
MAINTENANCE
AIRCRAFT CONFIGURATION CONTROL PANEL DIRECTIVE
PAGE 1 OF 1
CHANGE TITLE PAINT SCHEME/DECAL REVISION - T38
MODIFICATION EFFECTIVITY FLEET
ESTIMATED COST MANHOURS ENG 40 TECH | MATERIAL - | OTHER - | TOTAL EQUAL TO EXISTING COST
SCHEDULE AS SCHEDULED
PRIORITY
SAFETY OF FLIGHT
ORBITER SUPPORT
OTHER
WEIGHT
DISPOSITION AND DIRECTED ACTION
CHANGE TO BE IMPLEMENTED ONLY WHEN ALL INTERFACE ACTIONS HAVE BEEN AUTHORIZED
Revise the T38 paint scheme/decal location per 3809000 Rev G.
- Includes new NASA "SWOOSH" logo as an option.
- Reduces decal count from USAF ≈ 300.
CONCURRENCE:
MAINTENANCE DATE 6/1/93
OPERATIONS DATE 1 JUN 93
SAFETY DATE 1 JUN 93
QUALITY ASSURANCE DATE 1 Jun 93
ENGINEERING DATE 1 Jun 93
N/A
STA SIMULATION DATE
SUPPORT CONTRACTOR DATE 1 June 93
-NA-
SUPPORT CONTRACTOR DATE
RESPONSIBLE ENGINEER
R. M. PAYNE/NWASI
REVIEW REQUIREMENTS/RESPONSIBILITY
DESIGN REVIEW - | FRR/TRR - | FLIGHT TEST -
AUTHORIZATION
APPROVED ☑
APPRD WITH REVS ☐
DISAPPROVED ☐
SIGNATURE DATE 1 Jun 93
CHAIRMAN, CONFIGURATION CONTROL PANEL
CLOSEOUT CONCURRENCE:
AIRCRAFT MODIFICATION COMPLETE: 31 August 93 Date — Robert M Payne Engineer
DOCUMENTATION UPDATE COMPLETE: 31 aug 93 Date — Aircraft/ADAS Contractor Engineering
CCPD CLOSEOUT AUTHORIZATION: 31 aug 93 Date — Chairman, Configuration Control Panel
JSC Form 1692A (REV AUG 88) NASA-JSC

FIGURE 6-8.

The JSC work order that requested the changeover to the swoosh insignia for the NASA T-38 fleet. (Contributed by Robert M. Payne)

The Swoosh

Modarelli's meatball insignia has endured in an unaltered state since its reinstitution as NASA's official insignia in 1992. In the early 1990s, an unofficial modification to the meatball was conceived and implemented for NASA's fleet of research aircraft. Known as the "swoosh," the modified insignia is exhibited on today's NASA aircraft fleet.

In 1993, the Flight Crew Operations Directorate (FCOD) at the NASA Johnson Space Center (JSC) in Houston, Texas, was in charge of modifications and maintenance for the NASA-wide T-38 fleet. At the time, an avionics upgrade modification for the fleet was undertaken, followed by painting of the aircraft.[17]

14 "Modarelli Retires," *The Lewis News,* in-house newsletter, Volume 16 Number 2, 19 January 1979, p. 1.

15 "NACA Had a Unique Esprit de Corps," *The Lewis News,* in-house newsletter, 30 August 1991.

16 "Designer Reflects on Creation of NASA 'Meatball,'" *The Lewis News,* in-house newsletter, 3 July 1992, p. 1.

17 NASA retiree and former JSC pilot Roger Zwieg, telephone interview by Joseph Chambers, 20 March 2013.

FIGURE 6-9.

Two NASA T-38 aircraft at the Dryden Flight Research Center exhibit the swoosh tail insignia in 2007. (NASA ED07-0222-23)

Since the NASA worm had been retired in 1992, JSC was faced with the job of applying the meatball insignia to the tails of over 25 T-38s. As an alternative, the organization came up with a new, less complex version of the meatball for application to the T-38s. The designers of the swoosh emblem were Robert Payne and Thomas Grubbs of the FCOD, supported by JSC graphics personnel and Northrop engineering consultants.[18]

The new logo omitted the circular blue background and much of the details in the meatball, but retained the red supersonic wing. In addition, the letters of NASA and the orbiting spacecraft symbol were changed from white to blue.

The swoosh emblem appears on both sides of the vertical tail on NASA aircraft, with the "point" of the swoosh always to the rear. This configuration was chosen to accommodate the sweepback of the leading edge of the tail, but it results in a rather awkward appearance on the right side of the tail.

After JSC applied the swoosh emblem to the T-38 fleet, NASA's flight organization at Dryden Flight Research Center began to apply similar markings to its aircraft. Because the swoosh had not been formally approved by NASA Headquarters and no directive for specific markings existed, the Dryden applications differed slightly in style from those used by JSC. Implementation of the swoosh continued when the Langley and Lewis Research Centers also adopted the marking for their aircraft fleets. Today, the swoosh has been established as an option for applications to

18 E-mail communication with NASA JSC retiree Robert M. Payne, major participant in the development of the swoosh concept, 20 March 2013.

FIGURE 6-10.

FIGURE 6-11.

Details of the swoosh tail markings on NASA aircraft have been different. In this 2006 photograph of the NASA Langley OV-10 aircraft, the orbiting spacecraft insignia is red. Variations in color of the orbiting band on NASA aircraft have included red, black, and blue. (Contributed by Joseph Chambers)

This special meatball patch was designed and worn by NASA crews for deployments to the Middle East. (Contributed by Peter Merlin)

aircraft, with each organization responsible for its individual (although similar) marking.

Nowhere in the NASA Style Guide—the most current set of graphics standards available to NASA designers and artists—is the swoosh defined as an officially approved NASA emblem. The Style Guide still specifies the full-color meatball as the officially designated emblem for NASA aircraft.[19]

The flight research organizations of the NACA and NASA have historically been leaders in the institution and modification of insignia. Deployments of NASA

aircraft and aircrews to certain areas of the world have resulted in special markings and patches for flight suits. For example, deployments to the Middle East have led to modifications of the standard meatball patch. In one application, crews of the NASA WB-57F aircraft flying over Afghanistan had special "desert meatball" patches made that blended with their flight suits. In addition, NASA aircraft flying in those areas do not carry markings.[20]

19 Telephone interview with William B. Kluge, NASA graphic specialist who created Langley's version of the Swoosh marking, 5 March 2013.

20 E-mail correspondence with Robert Payne and Peter Merlin, 22 September 2014.

Applications of the Meatball

The NASA meatball insignia has three variations: a full-color insignia, a one-color insignia, and a one-color insignia with a white rule. Production requirements, media qualities, visibility, and proper usage determine the variation selected. The standards for the use of the NASA insignia and the NASA seal are in accordance with the Code of Federal Regulations, 14 CFR 1221, and the National Aeronautics and Space Act of 1958, as amended. NASA does not endorse any commercial product, activity, or service. Use of the NASA name, initials, or any associated emblem—including the NASA insignia, the retired NASA logotype, and the NASA seal—must be reviewed and approved by the Assistant Administrator for Public Affairs or a designee.[21]

An example of a commercial entity that successfully followed the approval process is the NASA Federal Credit Union (NASA FCU), which currently has over 100,000 members. In the 1970s, when the NASA logotype was implemented as the Agency's official insignia, the NASA FCU submitted a request to NASA for permission to use a blue version of the logotype within its emblem. The request was approved, and the symbol was "grandfathered" with continued approval after 1992 when the meatball was reinstituted as the NASA insignia.[22]

BUILDINGS AND EXHIBITS

During NASA's 40th anniversary celebration in 1998, the Agency painted the Vehicle Assembly Building at the Kennedy Space Center with the U.S. flag and the

21 NASA Style Guide, November 2006.

22 Electronic communication with Bert Ulrich, NASA Headquarters contact for Liaison Multimedia, by Joseph R. Chambers, 11 February 2015.

FIGURE 6-12.

The current logo of the NASA Federal Credit Union includes a unique blue version of the retired NASA logotype. (Courtesy of NASA FCU)

FIGURE 6-13.

The KSC Vehicle Assembly Building appears with the U.S. flag and meatball in 2011. Thousands of NASA Kennedy Space Center employees stand side-by-side to form a full-scale outline of a Space Shuttle orbiter outside the building. The unique photo opportunity was designed to honor the Space Shuttle Program's 30-year legacy and the people who contribute to safely processing, launching, and landing the vehicle. (NASA KSC-2011-2358)

meatball insignia.[23] During back-to-back hurricanes in 2004, the building lost nearly 850 of its panels that each measure 14 by 6 feet, and both the flag and meatball were damaged. The facility underwent a major repair and refurbishment project in 2007.

On 1 March 2014, NASA renamed its Dryden Flight Research Center the NASA Armstrong Flight Research Center (AFRC) in honor of the legendary astronaut Neil A. Armstrong. The Center continues to exhibit the NASA meatball insignia on virtually every building and research vehicle. The dominance of the meatball on display is common to all other Centers, with infrequent uses of the seal. An example of the seal can be seen at AFRC on the display of the HL-10 lifting body at the entrance to the Center.

AIRCRAFT

Even a cursory review of the markings and insignia carried by current-day NASA research and support aircraft reveals a wide array of practices and applications. The NACA maintained relatively disciplined markings on its aircraft. Today, however, NASA's aircraft are typically marked with program-specific logos determined by the resident research center.

---

23 "Restoring Old Glory and a Massive Meatball," available online at *http://www.nasa.gov/mission_pages/shuttle/behindscenes/vab_flag.html* (accessed 3 September 2014).

FIGURE 6-14.

The NASA seal is prominently exhibited on the HL-10 display at the entrance to the NASA Armstrong Flight Research Center. (NASA ED14-0081-128)

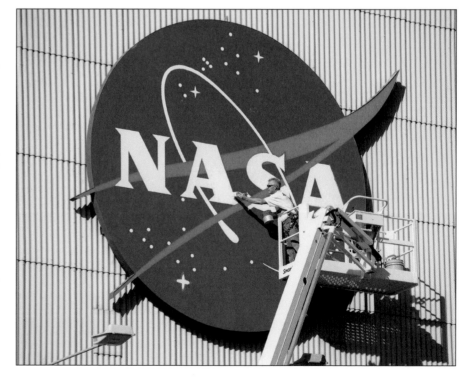

FIGURE 6-15.

The meatball on the Glenn Research Center's hangar gets a touch-up in 2006. (NASA C-2006-1777)

FIGURE 6-16.

NASA SR-71 and F-16XL aircraft fly in formation above Edwards, CA, during a sonic boom study in March 1995. The SR-71 (top) has a non-standard white tail band featuring the NASA logotype insignia, while the F-16XL has a stylized version of the swoosh insignia on its vertical tail. (NASA EC95-43024-2)

FIGURE 6-17.

The first of two unpiloted Global Hawk aircraft from the Armstrong Flight Research Center landed at the NASA Wallops Flight Facility in August 2014 to participate in research on hurricane formation and intensity. The meatball and swoosh insignia are carried by the vehicle, as are insignia of industry and government partners. (NASA WFF1-7967-0)

FIGURE 6-18.

NASA's Stratospheric Observatory for Infrared Astron-omy (SOFIA) flies with the sliding door over its tele-scope cavity fully open. Note the variations in markings, including the swoosh and insignia of partners. (NASA EC95-43024-2)

FIGURE 6-19.

NASA's NB-52B carrier aircraft rolls down a taxiway at Edwards Air Force Base with an unpiloted X-43A hyper-sonic scramjet vehicle mounted to a modified Pegasus booster rocket on a pylon under its right wing in 2001. The Pegasus launch vehicle exhibits the meatball, while the workhorse NB-52 still wears the gold NASA tail band of the 1960s. (NASA EC01-0079-3)

SPACE SHUTTLE

For over 30 years, NASA Space Shuttles accomplished "delivery truck" missions for a variety of objectives. The meatball insignia and the names and logos of NASA's partners typically appeared in the cargo bay or on the Shuttle's robotic arm.

Following the Space Shuttle Columbia disaster on 1 February 2003, NASA called on the Shuttle Discovery for a second Return-to-Flight mission. Discovery was launched on 26 July 2005 and delivered supplies to the ISS, landing on 9 August at Edwards. Discovery and its crew carried the meatball insignia on the mission. Once again, the insignia was on stage during intense media coverage of the event.

When the decision was made to terminate the Space Shuttle program in 2011 after 30 years of achievements and 135 missions, all active orbiters carried similar markings with the meatball insignia prominently displayed on the aft fuselage and left wing. As mentioned earlier, Shuttle Enterprise, a prototype vehicle, had retained its NASA logotype markings. Museums around the nation breathlessly awaited the outcome of the decision that would determine where the four vehicles would be displayed to preserve the legacy of NASA's Shuttle activities.

FIGURE 6-20.

The meatball insignia carried by the Shuttle Discovery on its left wing and aft fuselage are clearly visible in this dramatic picture of the launch of mission STS-120 in 2007. (NASA STS 120-S-047)

FIGURE 6-21.

Space Shuttles Enterprise, left, and Discovery meet nose-to-nose at the beginning of a transfer ceremony at the Smithsonian's Steven F. Udvar-Hazy Center on 19 April 2012 in Chantilly, Virginia. Discovery, with its NASA meatball insignia, took the place of Enterprise at the Center. (NASA/Smithsonian Institution)

FIGURE 6-22.

On 6 June 2012, Shuttle Enterprise is lifted off a barge and onto the Intrepid Sea, Air & Space Museum in New York City, where it is on display. Enterprise retained its worm insignia throughout its life. (NASA KSC)

FIGURE 6-23.

Space Shuttle Endeavour and the Shuttle Carrier Aircraft land at the Los Angeles International Airport on 12 September 2012, in preparation for entry into a museum display. Endeavour carried its meatball insignia into retirement and is on public display at the California Science Center in Los Angeles. (NASA ED12-0317-063)

FIGURE 6-24.

Shuttle Atlantis was the last to fly before the fleet was retired. It is on display at the NASA Kennedy Space Center Visitor Complex at Cape Canaveral, Florida. The open cargo bay doors obscure the meatball on the rear fuselage, but the insignia can be seen on the upper left wing. (NASA KSC 2013-2934)

SERVICE PINS

The meatball and the NASA seal appear on pins, medals, and plaques awarded for service to the Agency. A career service recognition award based on the meatball insignia is presented to all NASA civil service employees upon completion of 5 years of service. Certificates, emblems, and length of service mementos are presented at 35 years of service and for each 5-year interval thereafter. Employees with 40 years or more of federal service are eligible to receive certificates from the Administrator. Employees with 60 years or more of federal service are eligible to receive a letter from the President.[24]

DISTINGUISHED SERVICE MEDAL

The NASA Distinguished Service Medal is the only NASA medal with elements of the insignia or seal. It is NASA's highest form of recognition awarded to a government employee who, by distinguished service, ability, or vision, has personally contributed to NASA's advancement of U.S. interest. The individual's achievement or contribution must demonstrate a level of excellence that has made a profound or indelible impact on NASA mission success, and, therefore, the contribution is so extraordinary that other forms of recognition by NASA would be inadequate.

The first version of the medal (type I) featured the NASA seal and was issued from 1959 to 1961 to only three employees—John W. Crowley, NASA Director of Aeronautical and Space Research, and astronauts Alan B. Shepard, Jr., and Virgil I. "Gus" Grissom. As previously discussed, this version of the medal came under criticism from *Time* magazine after it was awarded to Shepard. The type I medal

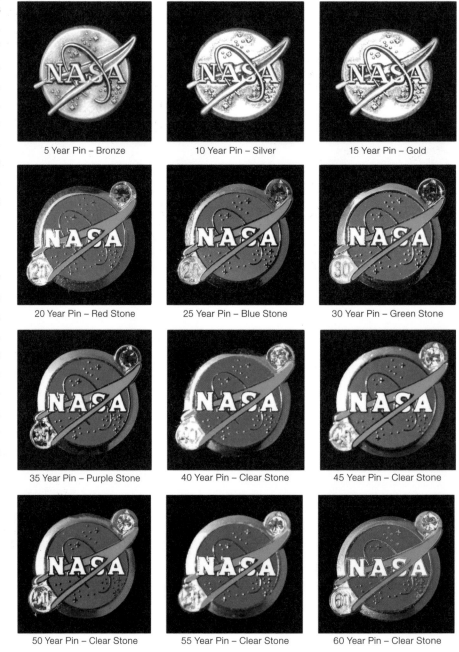

5 Year Pin – Bronze | 10 Year Pin – Silver | 15 Year Pin – Gold

20 Year Pin – Red Stone | 25 Year Pin – Blue Stone | 30 Year Pin – Green Stone

35 Year Pin – Purple Stone | 40 Year Pin – Clear Stone | 45 Year Pin – Clear Stone

50 Year Pin – Clear Stone | 55 Year Pin – Clear Stone | 60 Year Pin – Clear Stone

FIGURE 6-25.

NASA Service Pins are based on the meatball insignia. (NASA Employee Recognition and Awards Catalog 2014)

24 *https://searchpub.nssc.nasa.gov/servlet/sm.web.Fetch/ Appnd_H_-_Awards_Department_Catalog.pdf?rhid=1000 &did=1117274&type=released* (accessed 10 September 2014).

FIGURE 6-26.

Collection of NACA and NASA pins from the collection of former Ames employee Vernon Rogallo. (NASA Ames Artifact Collection number PP14.02)

FIGURE 6-27.

The NASA Distinguished Service Medal (type II) was created in 1961. (NASA)

was replaced with the type II version in 1961.[25] The new medal is a stylized version of the meatball logo, containing only the orbiting spacecraft and the supersonic wing emblems.

25 *http://www.omsa.org/photopost/showphoto.php?photo=2021* (accessed 10 September 2014)

FIGURE 6-28.

The NASA individual and group honor awards feature the NASA seal. (NASA)

HONOR AWARD CERTIFICATES

NASA also provides framed individual and group certificates for Agency and Center honor awards. The certificates feature the NASA seal and may be awarded to individuals or to groups of people. In addition to a framed award for group accomplishments, each individual group member receives a certificate to honor his or her accomplishment.

RETIREMENT PLAQUES

NASA employees retiring after 35 years of service receive a plaque in recognition of their career. The meatball is featured along with the employee's name and length of service.

FIGURE 6-29.

The meatball insignia is featured on retirement plaques for NASA employees. This blank plaque has been prepared for a future award. (NASA)

FIGURE 6-30.

Employees who worked for both the NACA and NASA received plaques with both logos. Calvin C. Berry retired as a supervisor of technicians after a 42-year career at Langley. (NASA Langley LHA)

INTERNAL NEWSLETTERS

NASA organizations disburse internal newsletters to employees to keep them informed and up to date on Agency news and the technical, social, and personal issues of the day. As the NASA insignia changed from the meatball to the logotype and back again, the publications changed in style and/or letterheads. In recent years, the newsletter delivery has become digital and computer-based, but the NASA insignia is still part of the graphics.

PUBLICATIONS

Over the years, the insignia has remained a part of the document. Current NASA publications include the meatball insignia on the cover.

FIGURE 6-31.

The letterhead of *The Researcher News,* Langley's in-house newsletter, has maintained a graphic of the NASA insignia through the years. Pictured are headers for (top to bottom) 1970, 1990, and 1996 reflecting the changes in the NASA insignia. (NASA)

FIGURE 6-32.

The appearance of NASA technical reports has changed considerably since the days of the NACA. For example, the cover of this 2007 technical publication does not cite the author's name and uses a Center-specific graphic. The NASA meatball is, however, retained on all reports. (NASA)

Miscellaneous Applications

The presence of the NASA meatball at virtually all NASA establishments and functions has become widespread, especially for special events and celebrations. Media coverage of NASA rarely occurs without a display of the meatball insignia, which is quickly recognized by the public. Recent special events have included renamings of NASA research centers and NASA-sponsored symposia and conferences.

FIGURE 6-33.

The NASA meatball dominates the backdrop, and the NASA seal appears on the speaker's podium on 7 May 1999 at a celebration of the renaming of the Lewis Research Center in honor of former astronaut John Glenn. The Center is now known as the NASA John H. Glenn Research Center at Lewis Field. (NASA C-99-1225)

FIGURE 6-34.

The meatball insignia was on display again on 2 March 2012, when 90-year-old John Glenn addressed a crowd of over 3,000 at Cleveland State University in celebration of being the first American to orbit Earth. (NASA C-2012-1298)

On 1 October 2008, NASA celebrated its 50th anniversary by issuing a special logo. The logo emphasized a futuristic design with the meatball insignia in company. The design, by Crabtree + Company, incorporates the Hubble Space Telescope image of the "grand design" spiral galaxy M81 located 11.6 million light years away in the constellation Ursa Major. The logo served as a centerpiece of Agency activities for the event, which included a special conference on the first 50 years of NASA.[26]

NASA's logos have inspired the young people of the nation via the Agency's educational programs, online activities, and interactive camps. The unbridled enthusiasm of children to absorb and embrace the NASA mission is stimulated by the iconic presence of the meatball. The meatball insignia now finds its way onto school supplies, backpacks, and apparel for children, as well as hats, jewelry, and other articles for adults. The items are available at numerous commercial locations, as well as at NASA gift shops located at NASA Headquarters and the NASA Centers.

Examples of youthful interest in the meatball abound, but one particularly interesting case involved the use of LEGO building blocks: "The NASA 'meatball' [took] shape brick by brick thanks to hundred of pieces of LEGO. The insignia was on display in the NASA exhibit at BrickFair 2011 near Washington, DC—an annual gathering of thousands of LEGO fans of all ages. The 2011 theme was 'NASA,' which

---

26 "NASA's First 50 Years: A Historical Perspective," conference held at NASA Headquarters, Washington, DC, 28–29 October 2008.

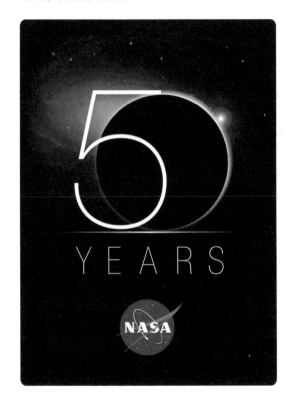

FIGURE 6-35.

The special logo adopted for the 50th Anniversary of NASA. (NASA C-2012-1298)

FIGURE 6-36.

The remarkable display of the meatball insignia at the 2011 BrickFair event. (NASA 580084-6712)

inspired many future (and current) design-and-builders to build aeronautics- or space-themed 'MOCs' (My Own Creation)."[27]

Several states, including Florida, Texas, and New York, have had space-themed automobile license plates. In 2002, a Virginia State Senator was approached by a Langley employee to provide advocacy for a special automobile vanity license plate to recognize the NASA Langley Research Center. Virginia has been very proactive in allowing different themes for vanity plates, and a positive response was obtained. Langley is currently the only NASA installation with vanity plates that can be purchased from the Division of Motor Vehicles.[28]

The NACA/NASA Centennial

The centennial anniversary of the NACA in 2015 celebrates a century of combined operations of the NACA and NASA. Major events and seminars carry a special logo designed in honor of the occasion. Emblematic of the linking of the two agencies, the logo uses interlocked letters to signify that union.

---

27 *http://www.nasa.gov/topics/aeronautics/features/brickfair1. html* (accessed 10 July 2015).

28 Langley employees Michael P. Finneran and Michael A. O'Hara initiated the action with Virginia State Senator Martin E. "Marty" Williams.

FIGURE 6-37.

The special Virginia vanity license plate created for the Langley Research Center. (NASA C-2012-1298)

FIGURE 6-38.

The special NASA logo for the centennial anniversary of NACA. (NASA via Tony Springer)

CHAPTER 7

Summary

This brief review of the conception, evolution, and applications of the seals, insignia, and logos of the NACA and NASA documents the history of these iconic symbols. The presence of these emblems during some of the nation's critical aeronautical and space-related activities has inspired widespread recognition, pride, and admiration for the agencies. The logos have been exhibited on buildings, aerospace vehicles, technical documents, flight suits, medals, and the personal attire of the NACA and NASA employees.

The review began with a discussion of the logos of the NACA. In its first 25 years of existence, the nation's first civil aeronautical research organization had become a world-class establishment with unique facilities, personnel, and technical capabilities in virtually every discipline associated with atmospheric flight. During the period, staff members embraced the creation of new emblems to distinguish the organization. The emergence of an informal NACA wings logo brought with it deep feelings of pride, team spirit, and international recognition. The original designer of the NACA emblem is unknown, but the design concept flourished despite the fact that the NACA did not have an official insignia until 1941.

The seal of the NACA was finally created and instituted as the Agency neared the end of its 43-year life in 1953. The insignia was internationally recognized and respected by the scientific community. As the Agency began to change its focus after WWII to the challenges of high-speed hypersonic flight and space, the launch of Sputnik by the Soviet Union led to the creation of a huge new organization that would bring its own legacy of logos and insignia.

The creation of NASA brought with it an attractive insignia and seal designed by Modarelli of the Lewis Research Center. The details of the designs included elements signifying the organization's missions in the fields of aeronautics, space exploration, and Earth sciences. The origin of the red wing used by Modarelli in the logos has been documented herein in detail. Modarelli's emblems were embraced by NASA in what would become an era of excitement and pride as the nation embarked on a major race for leadership in space research.

The logos became an intimate, ever-present component of activities as NASA achieved the necessary stepping-stones for travel to the Moon during the Mercury, Gemini, and Apollo programs. The meatball insignia, in particular, became a publicly admired symbol as the nation raced to meet a seemingly impossible schedule for a Moon landing imposed by a President with his eyes on the stars and the immediate Soviet threat. When Neil Armstrong and Buzz Aldrin set foot on the Moon on 20 July 1969, the meatball accompanied them on their thrilling lunar excursion.

The NASA meatball was replaced in 1975 by the artistic worm logotype insignia, much to the dismay of many NASA employees. Despite the logo controversy, the worm was carried during the early Shuttle missions and played a part in the advanced aircraft concepts conceived and evaluated by NASA and its industrial and military partners for 16 years. When new NASA Administrator Daniel Goldin retired the worm and brought back the meatball in 1992 as a reminder of the glory days of the Agency to improve the morale of employees, the logo was welcomed back by many veteran NASA employees.

Today, the NASA logos are instantly recognized throughout the world and continue to inspire and represent the world's leading civil Agency for research and development in aerospace technology. It is expected that the symbols will become even more visible as NASA celebrates the centennial anniversary of the NACA in 2015.

About the Authors

Joseph R. Chambers is an aviation consultant who lives in Yorktown, Virginia. He retired from the NASA Langley Research Center in 1998 after a 36-year career as a researcher and manager of military and civil aeronautics research activities. He began his career in 1962 as a member of the research staff of the Langley Full-Scale Tunnel, where he specialized in flight dynamics research on a variety of aerospace vehicles including V/STOL configurations, parawing vehicles, re-entry vehicles, and fighter-aircraft configurations. In 1974 he became the head of the Full-Scale Tunnel, the Langley 20-Foot Spin Tunnel, and outdoor free-flight and drop-model testing. In 1989 he became head of aircraft flight research at Langley in addition to his other responsibilities. In 1994 he was assigned to organize and manage a new group responsible for conducting systems-level analysis of the potential payoffs of NASA technologies and advanced aircraft concepts to help guide NASA research investments.

Chambers is the author of over 50 NASA technical reports and publications, including NASA Special Publications SP-514, *Patterns in the Sky,* on airflow condensation patterns for aircraft; SP-2000-4519, *Partners in Freedom,* on contributions of the Langley Research Center to U.S. military aircraft of the 1990s; SP-2003-4529, *Concept to Reality,* on contributions of Langley to U.S. civil aircraft of the 1990s; and SP-2005-4539, *Innovation in Flight,* on Langley research on advanced concepts for aeronautics.

He has written or contributed to several books for the NASA Aeronautics Research Mission Directorate, including SP-2009-575, *Modeling Flight,* on the development and application of dynamic free-flight models by the NACA and NASA. He was a contributor to SP-2010-570, *NASA's Contributions to Aeronautics.* His most recent publication is NASA SP-2014-614, *Cave of the Winds,* on the history of the Langley Full-Scale Wind Tunnel.

Chambers has made presentations on research and development programs to audiences as diverse as the von Karman Institute in Belgium and the annual Experimental Aircraft Association (EAA) Fly-In in Oshkosh, Wisconsin, and has consistently shown the ability to address a technical audience and the general public.

He has served as a representative of the United States on international committees and has given lectures on NASA's aeronautics programs in Japan, China, Australia, the United Kingdom, Canada, Italy, France, Germany, and Sweden.

Chambers received several of NASA's highest awards, including the Exceptional Service Medal, the Outstanding Leadership Medal, and the Public Service Medal. He also received the Arthur Flemming Award in 1975 as one of the 10 Most Outstanding Civil Servants for his management of NASA stall-spin research for military and civil aircraft. He has a bachelor of science degree from the Georgia Institute of Technology and a master of science degree from the Virginia Polytechnic Institute and State University (Virginia Tech).

Mark A. Chambers has conducted research and done technical writing in support of Langley's aeronautical and space research programs at the NASA Langley Research Center since 1991. While serving as a writer-editor in support of the Langley Office of Public Affairs and a contributor of technical articles to *The Researcher News*, Langley's in-house newsletter, he composed publications on aeronautics and space research including *The Birthplace of American Aviation: Milestones in Aviation and Aeronautical Research in the Mid-Atlantic Region*, an in-house reference guide for the Office of Public Affairs.

His career in support of Langley activities includes production of numerous articles, fact sheets, national news releases, historical summaries, and resource books. He currently serves as a senior technical writer for NCI Information Systems and has participated in writing and editing efforts in support of the Langley Aeronautics Research Directorate (ARD), the Langley Flight Projects Directorate (FPD), the Clouds and the Earth's Radiant Energy System (CERES) Program, and the Langley Ares I-X Systems Engineering and Integration (SE&I) Office.

His books include *From Research to Relevance: Significant Achievements in Aeronautical Research at Langley Research Center (1917–2002)* (NASA publication), *Engineering Test Pilot: The Exceptional Career of John P. "Jack" Reeder* (Virginia Aeronautical Historical Society publication), *Flight Research at NASA Langley Research Center* (Arcadia Publishing), *Radical Wings and Wind Tunnels: Advanced Concepts Tested at NASA Langley* (Specialty Press) (co-authored with his father, Joseph Chambers), *Naval Air Station Patuxent River* (Arcadia Publishing), and *Building the Supermarine Spitfire: Speed in the Skies* (The History Press Ltd., UK).

Chambers and his family live in Newport News, Virginia.

The NASA History Series

Reference Works, NASA SP-4000:

Grimwood, James M. *Project Mercury: A Chronology*. NASA SP-4001, 1963.

Grimwood, James M., and Barton C. Hacker, with Peter J. Vorzimmer. *Project Gemini Technology and Operations: A Chronology*. NASA SP-4002, 1969.

Link, Mae Mills. *Space Medicine in Project Mercury*. NASA SP-4003, 1965.

Astronautics and Aeronautics, 1963: Chronology of Science, Technology, and Policy. NASA SP-4004, 1964.

Astronautics and Aeronautics, 1964: Chronology of Science, Technology, and Policy. NASA SP-4005, 1965.

Astronautics and Aeronautics, 1965: Chronology of Science, Technology, and Policy. NASA SP-4006, 1966.

Astronautics and Aeronautics, 1966: Chronology of Science, Technology, and Policy. NASA SP-4007, 1967.

Astronautics and Aeronautics, 1967: Chronology of Science, Technology, and Policy. NASA SP-4008, 1968.

Ertel, Ivan D., and Mary Louise Morse. *The Apollo Spacecraft: A Chronology, Volume I, Through November 7, 1962*. NASA SP-4009, 1969.

Morse, Mary Louise, and Jean Kernahan Bays. *The Apollo Spacecraft: A Chronology, Volume II, November 8, 1962–September 30, 1964*. NASA SP-4009, 1973.

Brooks, Courtney G., and Ivan D. Ertel. *The Apollo Spacecraft: A Chronology, Volume III, October 1, 1964–January 20, 1966*. NASA SP-4009, 1973.

Ertel, Ivan D., and Roland W. Newkirk, with Courtney G. Brooks. *The Apollo Spacecraft: A Chronology, Volume IV, January 21, 1966–July 13, 1974*. NASA SP-4009, 1978.

Astronautics and Aeronautics, 1968: Chronology of Science, Technology, and Policy. NASA SP-4010, 1969.

Newkirk, Roland W., and Ivan D. Ertel, with Courtney G. Brooks. *Skylab: A Chronology*. NASA SP-4011, 1977.

Van Nimmen, Jane, and Leonard C. Bruno, with Robert L. Rosholt. *NASA Historical Data Book, Volume I: NASA Resources, 1958–1968*. NASA SP-4012, 1976; rep. ed. 1988.

Ezell, Linda Neuman. *NASA Historical Data Book, Volume II: Programs and Projects, 1958–1968*. NASA SP-4012, 1988.

Ezell, Linda Neuman. *NASA Historical Data Book, Volume III: Programs and Projects, 1969–1978*. NASA SP-4012, 1988.

Gawdiak, Ihor, with Helen Fedor. *NASA Historical Data Book, Volume IV: NASA Resources, 1969–1978*. NASA SP-4012, 1994.

Rumerman, Judy A. *NASA Historical Data Book, Volume V: NASA Launch Systems, Space Transportation, Human Spaceflight, and Space Science, 1979–1988*. NASA SP-4012, 1999.

Rumerman, Judy A. *NASA Historical Data Book, Volume VI: NASA Space Applications, Aeronautics and Space Research and Technology, Tracking and Data Acquisition/Support Operations,*

Commercial Programs, and Resources, 1979–1988. NASA SP-4012, 1999.

Rumerman, Judy A. *NASA Historical Data Book, Volume VII: NASA Launch Systems, Space Transportation, Human Spaceflight, and Space Science, 1989–1998.* NASA SP-2009-4012, 2009.

Rumerman, Judy A. *NASA Historical Data Book, Volume VIII: NASA Earth Science and Space Applications, Aeronautics, Technology, and Exploration, Tracking and Data Acquisition/ Space Operations, Facilities and Resources, 1989–1998.* NASA SP-2012-4012, 2012.

No SP-4013.

Astronautics and Aeronautics, 1969: Chronology of Science, Technology, and Policy. NASA SP-4014, 1970.

Astronautics and Aeronautics, 1970: Chronology of Science, Technology, and Policy. NASA SP-4015, 1972.

Astronautics and Aeronautics, 1971: Chronology of Science, Technology, and Policy. NASA SP-4016, 1972.

Astronautics and Aeronautics, 1972: Chronology of Science, Technology, and Policy. NASA SP-4017, 1974.

Astronautics and Aeronautics, 1973: Chronology of Science, Technology, and Policy. NASA SP-4018, 1975.

Astronautics and Aeronautics, 1974: Chronology of Science, Technology, and Policy. NASA SP-4019, 1977.

Astronautics and Aeronautics, 1975: Chronology of Science, Technology, and Policy. NASA SP-4020, 1979.

Astronautics and Aeronautics, 1976: Chronology of Science, Technology, and Policy. NASA SP-4021, 1984.

Astronautics and Aeronautics, 1977: Chronology of Science, Technology, and Policy. NASA SP-4022, 1986.

Astronautics and Aeronautics, 1978: Chronology of Science, Technology, and Policy. NASA SP-4023, 1986.

Astronautics and Aeronautics, 1979–1984: Chronology of Science, Technology, and Policy. NASA SP-4024, 1988.

Astronautics and Aeronautics, 1985: Chronology of Science, Technology, and Policy. NASA SP-4025, 1990.

Noordung, Hermann. *The Problem of Space Travel: The Rocket Motor.* Edited by Ernst Stuhlinger and J. D. Hunley, with Jennifer Garland. NASA SP-4026, 1995.

Gawdiak, Ihor Y., Ramon J. Miro, and Sam Stueland. *Astronautics and Aeronautics, 1986–1990: A Chronology.* NASA SP-4027, 1997.

Gawdiak, Ihor Y., and Charles Shetland. *Astronautics and Aeronautics, 1991–1995: A Chronology.* NASA SP-2000-4028, 2000.

Orloff, Richard W. *Apollo by the Numbers: A Statistical Reference.* NASA SP-2000-4029, 2000.

Lewis, Marieke, and Ryan Swanson. *Astronautics and Aeronautics: A Chronology, 1996–2000.* NASA SP-2009-4030, 2009.

Ivey, William Noel, and Marieke Lewis. *Astronautics and Aeronautics: A Chronology, 2001–2005.* NASA SP-2010-4031, 2010.

Buchalter, Alice R., and William Noel Ivey. *Astronautics and Aeronautics: A Chronology, 2006.* NASA SP-2011-4032, 2010.

Lewis, Marieke. *Astronautics and Aeronautics: A Chronology, 2007.* NASA SP-2011-4033, 2011.

Lewis, Marieke. *Astronautics and Aeronautics: A Chronology, 2008.* NASA SP-2012-4034, 2012.

Lewis, Marieke. *Astronautics and Aeronautics: A Chronology, 2009.* NASA SP-2012-4035, 2012.

Flattery, Meaghan. *Astronautics and Aeronautics: A Chronology,* 2010. NASA SP-2013-4037, 2014.

Management Histories, NASA SP-4100:

Rosholt, Robert L. *An Administrative History of NASA, 1958–1963.* NASA SP-4101, 1966.

Levine, Arnold S. *Managing NASA in the Apollo Era.* NASA SP-4102, 1982.

Roland, Alex. *Model Research: The National Advisory Committee for Aeronautics, 1915–1958.* NASA SP-4103, 1985.

Fries, Sylvia D. *NASA Engineers and the Age of Apollo.* NASA SP-4104, 1992.

Glennan, T. Keith. *The Birth of NASA: The Diary of T. Keith Glennan.* Edited by J. D. Hunley. NASA SP-4105, 1993.

Seamans, Robert C. *Aiming at Targets: The Autobiography of Robert C. Seamans.* NASA SP-4106, 1996.

Garber, Stephen J., ed. *Looking Backward, Looking Forward: Forty Years of Human Spaceflight Symposium*. NASA SP-2002-4107, 2002.

Mallick, Donald L., with Peter W. Merlin. *The Smell of Kerosene: A Test Pilot's Odyssey*. NASA SP-4108, 2003.

Iliff, Kenneth W., and Curtis L. Peebles. *From Runway to Orbit: Reflections of a NASA Engineer*. NASA SP-2004-4109, 2004.

Chertok, Boris. *Rockets and People, Volume I*. NASA SP-2005-4110, 2005.

Chertok, Boris. *Rockets and People: Creating a Rocket Industry, Volume II*. NASA SP-2006-4110, 2006.

Chertok, Boris. *Rockets and People: Hot Days of the Cold War, Volume III*. NASA SP-2009-4110, 2009.

Chertok, Boris. *Rockets and People: The Moon Race, Volume IV*. NASA SP-2011-4110, 2011.

Laufer, Alexander, Todd Post, and Edward Hoffman. *Shared Voyage: Learning and Unlearning from Remarkable Projects*. NASA SP-2005-4111, 2005.

Dawson, Virginia P., and Mark D. Bowles. *Realizing the Dream of Flight: Biographical Essays in Honor of the Centennial of Flight, 1903–2003*. NASA SP-2005-4112, 2005.

Mudgway, Douglas J. *William H. Pickering: America's Deep Space Pioneer*. NASA SP-2008-4113, 2008.

Wright, Rebecca, Sandra Johnson, and Steven J. Dick. *NASA at 50: Interviews with NASA's Senior Leadership*. NASA SP-2012-4114, 2012.

Project Histories, NASA SP-4200:

Swenson, Loyd S., Jr., James M. Grimwood, and Charles C. Alexander. *This New Ocean: A History of Project Mercury*. NASA SP-4201, 1966; rep. ed. 1999.

Green, Constance McLaughlin, and Milton Lomask. *Vanguard: A History*. NASA SP-4202, 1970; rep. ed. Smithsonian Institution Press, 1971.

Hacker, Barton C., and James M. Grimwood. *On the Shoulders of Titans: A History of Project Gemini*. NASA SP-4203, 1977; rep. ed. 2002.

Benson, Charles D., and William Barnaby Faherty. *Moonport: A History of Apollo Launch Facilities and Operations*. NASA SP-4204, 1978.

Brooks, Courtney G., James M. Grimwood, and Loyd S. Swenson, Jr. *Chariots for Apollo: A History of Manned Lunar Spacecraft*. NASA SP-4205, 1979.

Bilstein, Roger E. *Stages to Saturn: A Technological History of the Apollo/Saturn Launch Vehicles*. NASA SP-4206, 1980 and 1996.

No SP-4207.

Compton, W. David, and Charles D. Benson. *Living and Working in Space: A History of Skylab*. NASA SP-4208, 1983.

Ezell, Edward Clinton, and Linda Neuman Ezell. *The Partnership: A History of the Apollo-Soyuz Test Project*. NASA SP-4209, 1978.

Hall, R. Cargill. *Lunar Impact: A History of Project Ranger*. NASA SP-4210, 1977.

Newell, Homer E. *Beyond the Atmosphere: Early Years of Space Science*. NASA SP-4211, 1980.

Ezell, Edward Clinton, and Linda Neuman Ezell. *On Mars: Exploration of the Red Planet, 1958–1978*. NASA SP-4212, 1984.

Pitts, John A. *The Human Factor: Biomedicine in the Manned Space Program to 1980*. NASA SP-4213, 1985.

Compton, W. David. *Where No Man Has Gone Before: A History of Apollo Lunar Exploration Missions*. NASA SP-4214, 1989.

Naugle, John E. *First Among Equals: The Selection of NASA Space Science Experiments*. NASA SP-4215, 1991.

Wallace, Lane E. *Airborne Trailblazer: Two Decades with NASA Langley's 737 Flying Laboratory*. NASA SP-4216, 1994.

Butrica, Andrew J., ed. *Beyond the Ionosphere: Fifty Years of Satellite Communications*. NASA SP-4217, 1997.

Butrica, Andrew J. *To See the Unseen: A History of Planetary Radar Astronomy*. NASA SP-4218, 1996.

Mack, Pamela E., ed. *From Engineering Science to Big Science: The NACA and NASA Collier Trophy Research Project Winners*. NASA SP-4219, 1998.

Reed, R. Dale. *Wingless Flight: The Lifting Body Story.* NASA SP-4220, 1998.

Heppenheimer, T. A. *The Space Shuttle Decision: NASA's Search for a Reusable Space Vehicle.* NASA SP-4221, 1999.

Hunley, J. D., ed. *Toward Mach 2: The Douglas D-558 Program.* NASA SP-4222, 1999.

Swanson, Glen E., ed. *"Before This Decade Is Out…" Personal Reflections on the Apollo Program.* NASA SP-4223, 1999.

Tomayko, James E. *Computers Take Flight: A History of NASA's Pioneering Digital Fly-By-Wire Project.* NASA SP-4224, 2000.

Morgan, Clay. *Shuttle-Mir: The United States and Russia Share History's Highest Stage.* NASA SP-2001-4225, 2001.

Leary, William M. *"We Freeze to Please": A History of NASA's Icing Research Tunnel and the Quest for Safety.* NASA SP-2002-4226, 2002.

Mudgway, Douglas J. *Uplink-Downlink: A History of the Deep Space Network, 1957–1997.* NASA SP-2001-4227, 2001.

No SP-4228 or SP-4229.

Dawson, Virginia P., and Mark D. Bowles. *Taming Liquid Hydrogen: The Centaur Upper Stage Rocket, 1958–2002.* NASA SP-2004-4230, 2004.

Meltzer, Michael. *Mission to Jupiter: A History of the Galileo Project.* NASA SP-2007-4231, 2007.

Heppenheimer, T. A. *Facing the Heat Barrier: A History of Hypersonics.* NASA SP-2007-4232, 2007.

Tsiao, Sunny. *"Read You Loud and Clear!" The Story of NASA's Spaceflight Tracking and Data Network.* NASA SP-2007-4233, 2007.

Meltzer, Michael. *When Biospheres Collide: A History of NASA's Planetary Protection Programs.* NASA SP-2011-4234, 2011.

Center Histories, NASA SP-4300:

Rosenthal, Alfred. *Venture into Space: Early Years of Goddard Space Flight Center.* NASA SP-4301, 1985.

Hartman, Edwin P. *Adventures in Research: A History of Ames Research Center, 1940–1965.* NASA SP-4302, 1970.

Hallion, Richard P. *On the Frontier: Flight Research at Dryden, 1946–1981.* NASA SP-4303, 1984.

Muenger, Elizabeth A. *Searching the Horizon: A History of Ames Research Center, 1940–1976.* NASA SP-4304, 1985.

Hansen, James R. *Engineer in Charge: A History of the Langley Aeronautical Laboratory, 1917–1958.* NASA SP-4305, 1987.

Dawson, Virginia P. *Engines and Innovation: Lewis Laboratory and American Propulsion Technology.* NASA SP-4306, 1991.

Dethloff, Henry C. *"Suddenly Tomorrow Came…": A History of the Johnson Space Center, 1957–1990.* NASA SP-4307, 1993.

Hansen, James R. *Spaceflight Revolution: NASA Langley Research Center from Sputnik to Apollo.* NASA SP-4308, 1995.

Wallace, Lane E. *Flights of Discovery: An Illustrated History of the Dryden Flight Research Center.* NASA SP-4309, 1996.

Herring, Mack R. *Way Station to Space: A History of the John C. Stennis Space Center.* NASA SP-4310, 1997.

Wallace, Harold D., Jr. *Wallops Station and the Creation of an American Space Program.* NASA SP-4311, 1997.

Wallace, Lane E. *Dreams, Hopes, Realities. NASA's Goddard Space Flight Center: The First Forty Years.* NASA SP-4312, 1999.

Dunar, Andrew J., and Stephen P. Waring. *Power to Explore: A History of Marshall Space Flight Center, 1960–1990.* NASA SP-4313, 1999.

Bugos, Glenn E. *Atmosphere of Freedom: Sixty Years at the NASA Ames Research Center.* NASA SP-2000-4314, 2000.

Bugos, Glenn E. *Atmosphere of Freedom: Seventy Years at the NASA Ames Research Center.* NASA SP-2010-4314, 2010. Revised version of NASA SP-2000-4314.

Bugos, Glenn E. *Atmosphere of Freedom: Seventy Five Years at the NASA Ames Research Center.* NASA SP-2014-4314, 2014. Revised version of NASA SP-2000-4314.

No SP-4315.

Schultz, James. *Crafting Flight: Aircraft Pioneers and the Contributions of the Men and Women of NASA Langley Research Center.* NASA SP-2003-4316, 2003.

Bowles, Mark D. *Science in Flux: NASA's Nuclear Program at Plum Brook Station, 1955–2005*. NASA SP-2006-4317, 2006.

Wallace, Lane E. *Flights of Discovery: An Illustrated History of the Dryden Flight Research Center*. NASA SP-2007-4318, 2007. Revised version of NASA SP-4309.

Arrighi, Robert S. *Revolutionary Atmosphere: The Story of the Altitude Wind Tunnel and the Space Power Chambers*. NASA SP-2010-4319, 2010.

General Histories, NASA SP-4400:

Corliss, William R. *NASA Sounding Rockets, 1958–1968: A Historical Summary*. NASA SP-4401, 1971.

Wells, Helen T., Susan H. Whiteley, and Carrie Karegeannes. *Origins of NASA Names*. NASA SP-4402, 1976.

Anderson, Frank W., Jr. *Orders of Magnitude: A History of NACA and NASA, 1915–1980*. NASA SP-4403, 1981.

Sloop, John L. *Liquid Hydrogen as a Propulsion Fuel, 1945–1959*. NASA SP-4404, 1978.

Roland, Alex. *A Spacefaring People: Perspectives on Early Spaceflight*. NASA SP-4405, 1985.

Bilstein, Roger E. *Orders of Magnitude: A History of the NACA and NASA, 1915–1990*. NASA SP-4406, 1989.

Logsdon, John M., ed., with Linda J. Lear, Jannelle Warren Findley, Ray A. Williamson, and Dwayne A. Day. *Exploring the Unknown: Selected Documents in the History of the U.S. Civil Space Program, Volume I: Organizing for Exploration*. NASA SP-4407, 1995.

Logsdon, John M., ed., with Dwayne A. Day and Roger D. Launius. *Exploring the Unknown: Selected Documents in the History of the U.S. Civil Space Program, Volume II: External Relationships*. NASA SP-4407, 1996.

Logsdon, John M., ed., with Roger D. Launius, David H. Onkst, and Stephen J. Garber. *Exploring the Unknown: Selected Documents in the History of the U.S. Civil Space Program, Volume III: Using Space*. NASA SP-4407, 1998.

Logsdon, John M., ed., with Ray A. Williamson, Roger D. Launius, Russell J. Acker, Stephen J. Garber, and Jonathan L. Friedman. *Exploring the Unknown: Selected Documents in the History of the U.S. Civil Space Program, Volume IV: Accessing Space*. NASA SP-4407, 1999.

Logsdon, John M., ed., with Amy Paige Snyder, Roger D. Launius, Stephen J. Garber, and Regan Anne Newport. *Exploring the Unknown: Selected Documents in the History of the U.S. Civil Space Program, Volume V: Exploring the Cosmos*. NASA SP-2001-4407, 2001.

Logsdon, John M., ed., with Stephen J. Garber, Roger D. Launius, and Ray A. Williamson. *Exploring the Unknown: Selected Documents in the History of the U.S. Civil Space Program, Volume VI: Space and Earth Science*. NASA SP-2004-4407, 2004.

Logsdon, John M., ed., with Roger D. Launius. *Exploring the Unknown: Selected Documents in the History of the U.S. Civil Space Program, Volume VII: Human Spaceflight: Projects Mercury, Gemini, and Apollo*. NASA SP-2008-4407, 2008.

Siddiqi, Asif A., *Challenge to Apollo: The Soviet Union and the Space Race, 1945–1974*. NASA SP-2000-4408, 2000.

Hansen, James R., ed. *The Wind and Beyond: Journey into the History of Aerodynamics in America, Volume 1: The Ascent of the Airplane*. NASA SP-2003-4409, 2003.

Hansen, James R., ed. *The Wind and Beyond: Journey into the History of Aerodynamics in America, Volume 2: Reinventing the Airplane*. NASA SP-2007-4409, 2007.

Hogan, Thor. *Mars Wars: The Rise and Fall of the Space Exploration Initiative*. NASA SP-2007-4410, 2007.

Vakoch, Douglas A., ed. *Psychology of Space Exploration: Contemporary Research in Historical Perspective*. NASA SP-2011-4411, 2011.

Ferguson, Robert G., *NASA's First A: Aeronautics from 1958 to 2008*. NASA SP-2012-4412, 2013.

Vakoch, Douglas A., ed. *Archaeology, Anthropology, and Interstellar Communication*. NASA SP-2013-4413, 2014.

Monographs in Aerospace History, NASA SP-4500:

Launius, Roger D., and Aaron K. Gillette, comps. *Toward a History of the Space Shuttle: An Annotated Bibliography*. Monographs in Aerospace History, No. 1, 1992.

Launius, Roger D., and J. D. Hunley, comps. *An Annotated Bibliography of the Apollo Program*. Monographs in Aerospace History, No. 2, 1994.

Launius, Roger D. *Apollo: A Retrospective Analysis*. Monographs in Aerospace History, No. 3, 1994.

Hansen, James R. *Enchanted Rendezvous: John C. Houbolt and the Genesis of the Lunar-Orbit Rendezvous Concept*. Monographs in Aerospace History, No. 4, 1995.

Gorn, Michael H. *Hugh L. Dryden's Career in Aviation and Space*. Monographs in Aerospace History, No. 5, 1996.

Powers, Sheryll Goecke. *Women in Flight Research at NASA Dryden Flight Research Center from 1946 to 1995*. Monographs in Aerospace History, No. 6, 1997.

Portree, David S. F., and Robert C. Trevino. *Walking to Olympus: An EVA Chronology*. Monographs in Aerospace History, No. 7, 1997.

Logsdon, John M., moderator. *Legislative Origins of the National Aeronautics and Space Act of 1958: Proceedings of an Oral History Workshop*. Monographs in Aerospace History, No. 8, 1998.

Rumerman, Judy A., comp. *U.S. Human Spaceflight: A Record of Achievement, 1961–1998*. Monographs in Aerospace History, No. 9, 1998.

Portree, David S. F. *NASA's Origins and the Dawn of the Space Age*. Monographs in Aerospace History, No. 10, 1998.

Logsdon, John M. *Together in Orbit: The Origins of International Cooperation in the Space Station*. Monographs in Aerospace History, No. 11, 1998.

Phillips, W. Hewitt. *Journey in Aeronautical Research: A Career at NASA Langley Research Center*. Monographs in Aerospace History, No. 12, 1998.

Braslow, Albert L. *A History of Suction-Type Laminar-Flow Control with Emphasis on Flight Research*. Monographs in Aerospace History, No. 13, 1999.

Logsdon, John M., moderator. *Managing the Moon Program: Lessons Learned from Apollo*. Monographs in Aerospace History, No. 14, 1999.

Perminov, V. G. *The Difficult Road to Mars: A Brief History of Mars Exploration in the Soviet Union*. Monographs in Aerospace History, No. 15, 1999.

Tucker, Tom. *Touchdown: The Development of Propulsion Controlled Aircraft at NASA Dryden*. Monographs in Aerospace History, No. 16, 1999.

Maisel, Martin, Demo J. Giulanetti, and Daniel C. Dugan. *The History of the XV-15 Tilt Rotor Research Aircraft: From Concept to Flight*. Monographs in Aerospace History, No. 17, 2000. NASA SP-2000-4517.

Jenkins, Dennis R. *Hypersonics Before the Shuttle: A Concise History of the X-15 Research Airplane*. Monographs in Aerospace History, No. 18, 2000. NASA SP-2000-4518.

Chambers, Joseph R. *Partners in Freedom: Contributions of the Langley Research Center to U.S. Military Aircraft of the 1990s*. Monographs in Aerospace History, No. 19, 2000. NASA SP-2000-4519.

Waltman, Gene L. *Black Magic and Gremlins: Analog Flight Simulations at NASA's Flight Research Center*. Monographs in Aerospace History, No. 20, 2000. NASA SP-2000-4520.

Portree, David S. F. *Humans to Mars: Fifty Years of Mission Planning, 1950–2000*. Monographs in Aerospace History, No. 21, 2001. NASA SP-2001-4521.

Thompson, Milton O., with J. D. Hunley. *Flight Research: Problems Encountered and What They Should Teach Us*. Monographs in Aerospace History, No. 22, 2001. NASA SP-2001-4522.

Tucker, Tom. *The Eclipse Project*. Monographs in Aerospace History, No. 23, 2001. NASA SP-2001-4523.

Siddiqi, Asif A. *Deep Space Chronicle: A Chronology of Deep Space and Planetary Probes, 1958–2000*. Monographs in Aerospace History, No. 24, 2002. NASA SP-2002-4524.

Merlin, Peter W. *Mach 3+: NASA/USAF YF-12 Flight Research, 1969–1979*. Monographs in Aerospace History, No. 25, 2001. NASA SP-2001-4525.

Anderson, Seth B. *Memoirs of an Aeronautical Engineer: Flight Tests at Ames Research Center: 1940–1970.* Monographs in Aerospace History, No. 26, 2002. NASA SP-2002-4526.

Renstrom, Arthur G. *Wilbur and Orville Wright: A Bibliography Commemorating the One-Hundredth Anniversary of the First Powered Flight on December 17, 1903.* Monographs in Aerospace History, No. 27, 2002. NASA SP-2002-4527.

No monograph 28.

Chambers, Joseph R. *Concept to Reality: Contributions of the NASA Langley Research Center to U.S. Civil Aircraft of the 1990s.* Monographs in Aerospace History, No. 29, 2003. NASA SP-2003-4529.

Peebles, Curtis, ed. *The Spoken Word: Recollections of Dryden History, The Early Years.* Monographs in Aerospace History, No. 30, 2003. NASA SP-2003-4530.

Jenkins, Dennis R., Tony Landis, and Jay Miller. *American X-Vehicles: An Inventory—X-1 to X-50.* Monographs in Aerospace History, No. 31, 2003. NASA SP-2003-4531.

Renstrom, Arthur G. *Wilbur and Orville Wright: A Chronology Commemorating the One-Hundredth Anniversary of the First Powered Flight on December 17, 1903.* Monographs in Aerospace History, No. 32, 2003. NASA SP-2003-4532.

Bowles, Mark D., and Robert S. Arrighi. *NASA's Nuclear Frontier: The Plum Brook Research Reactor.* Monographs in Aerospace History, No. 33, 2004. NASA SP-2004-4533.

Wallace, Lane, and Christian Gelzer. *Nose Up: High Angle-of-Attack and Thrust Vectoring Research at NASA Dryden, 1979–2001.* Monographs in Aerospace History, No. 34, 2009. NASA SP-2009-4534.

Matranga, Gene J., C. Wayne Ottinger, Calvin R. Jarvis, and D. Christian Gelzer. *Unconventional, Contrary, and Ugly: The Lunar Landing Research Vehicle.* Monographs in Aerospace History, No. 35, 2006. NASA SP-2004-4535.

McCurdy, Howard E. *Low-Cost Innovation in Spaceflight: The History of the Near Earth Asteroid Rendezvous (NEAR) Mission.* Monographs in Aerospace History, No. 36, 2005. NASA SP-2005-4536.

Seamans, Robert C., Jr. *Project Apollo: The Tough Decisions.* Monographs in Aerospace History, No. 37, 2005. NASA SP-2005-4537.

Lambright, W. Henry. *NASA and the Environment: The Case of Ozone Depletion.* Monographs in Aerospace History, No. 38, 2005. NASA SP-2005-4538.

Chambers, Joseph R. *Innovation in Flight: Research of the NASA Langley Research Center on Revolutionary Advanced Concepts for Aeronautics.* Monographs in Aerospace History, No. 39, 2005. NASA SP-2005-4539.

Phillips, W. Hewitt. *Journey into Space Research: Continuation of a Career at NASA Langley Research Center.* Monographs in Aerospace History, No. 40, 2005. NASA SP-2005-4540.

Rumerman, Judy A., Chris Gamble, and Gabriel Okolski, comps. *U.S. Human Spaceflight: A Record of Achievement, 1961–2006.* Monographs in Aerospace History, No. 41, 2007. NASA SP-2007-4541.

Peebles, Curtis. *The Spoken Word: Recollections of Dryden History Beyond the Sky.* Monographs in Aerospace History, No. 42, 2011. NASA SP-2011-4542.

Dick, Steven J., Stephen J. Garber, and Jane H. Odom. *Research in NASA History.* Monographs in Aerospace History, No. 43, 2009. NASA SP-2009-4543.

Merlin, Peter W. *Ikhana: Unmanned Aircraft System Western States Fire Missions.* Monographs in Aerospace History, No. 44, 2009. NASA SP-2009-4544.

Fisher, Steven C., and Shamim A. Rahman. *Remembering the Giants: Apollo Rocket Propulsion Development.* Monographs in Aerospace History, No. 45, 2009. NASA SP-2009-4545.

Gelzer, Christian. *Fairing Well: From Shoebox to Bat Truck and Beyond, Aerodynamic Truck Research at NASA's Dryden Flight Research Center.* Monographs in Aerospace History, No. 46, 2011. NASA SP-2011-4546.

Arrighi, Robert. *Pursuit of Power: NASA's Propulsion Systems Laboratory No. 1 and 2.* Monographs in Aerospace History, No. 48, 2012. NASA SP-2012-4548.

Goodrich, Malinda K., Alice R. Buchalter, and Patrick M. Miller, comps. *Toward a History of the Space Shuttle: An Annotated*

Bibliography, Part 2 (1992–2011). Monographs in Aerospace History, No. 49, 2012. NASA SP-2012-4549.

Gelzer, Christian. *The Spoken Word III: Recollections of Dryden History; The Shuttle Years.* Monographs in Aerospace History, No. 52, 2013. NASA SP-2013-4552.

Ross, James C. *NASA Photo One.* Monographs in Aerospace History, No. 53, 2013. NASA SP-2013-4553.

Launius, Roger D. *Historical Analogs for the Stimulation of Space Commerce.* Monographs in Aerospace History, No 54, 2014. NASA SP-2014-4554.

Buchalter, Alice R., and Patrick M. Miller, comps. *The National Advisory Committee for Aeronautics: An Annotated Bibliography.* Monographs in Aerospace History, No. 55, 2014. NASA SP-2014-4555.

Electronic Media, NASA SP-4600:

Remembering Apollo 11: The 30th Anniversary Data Archive CD-ROM. NASA SP-4601, 1999.

Remembering Apollo 11: The 35th Anniversary Data Archive CD-ROM. NASA SP-2004-4601, 2004. This is an update of the 1999 edition.

The Mission Transcript Collection: U.S. Human Spaceflight Missions from Mercury Redstone 3 to Apollo 17. NASA SP-2000-4602, 2001.

Shuttle-Mir: The United States and Russia Share History's Highest Stage. NASA SP-2001-4603, 2002.

U.S. Centennial of Flight Commission Presents Born of Dreams— Inspired by Freedom. NASA SP-2004-4604, 2004.

Of Ashes and Atoms: A Documentary on the NASA Plum Brook Reactor Facility. NASA SP-2005-4605, 2005.

Taming Liquid Hydrogen: The Centaur Upper Stage Rocket Interactive CD-ROM. NASA SP-2004-4606, 2004.

Fueling Space Exploration: The History of NASA's Rocket Engine Test Facility DVD. NASA SP-2005-4607, 2005.

Altitude Wind Tunnel at NASA Glenn Research Center: An Interactive History CD-ROM. NASA SP-2008-4608, 2008.

A Tunnel Through Time: The History of NASA's Altitude Wind Tunnel. NASA SP-2010-4609, 2010.

Conference Proceedings, NASA SP-4700:

Dick, Steven J., and Keith Cowing, eds. *Risk and Exploration: Earth, Sea and the Stars.* NASA SP-2005-4701, 2005.

Dick, Steven J., and Roger D. Launius. *Critical Issues in the History of Spaceflight.* NASA SP-2006-4702, 2006.

Dick, Steven J., ed. *Remembering the Space Age: Proceedings of the 50th Anniversary Conference.* NASA SP-2008-4703, 2008.

Dick, Steven J., ed. *NASA's First 50 Years: Historical Perspectives.* NASA SP-2010-4704, 2010.

Societal Impact, NASA SP-4800:

Dick, Steven J., and Roger D. Launius. *Societal Impact of Spaceflight.* NASA SP-2007-4801, 2007.

Dick, Steven J., and Mark L. Lupisella. *Cosmos and Culture: Cultural Evolution in a Cosmic Context.* NASA SP-2009-4802, 2009.

Index